The publisher gratefully acknowledges the generous support of the Valerie Barth and Peter Booth Wiley Endowment Fund in History of the University of California Press Foundation.

*Flame and Fortune
in the American West*

CRITICAL ENVIRONMENTS: NATURE, SCIENCE, AND POLITICS

Edited by Julie Guthman, Jake Kosek, and Rebecca Lave

The Critical Environments series publishes books that explore the political forms of life and the ecologies that emerge from histories of capitalism, militarism, racism, colonialism, and more.

1. *Flame and Fortune in the American West: Urban Development, Environmental Change, and the Great Oakland Hills Fire,* by Gregory L. Simon
2. *Germ Wars: The Politics of Microbes and America's Landscape of Fear,* by Melanie Armstrong

Flame and Fortune in the American West

URBAN DEVELOPMENT, ENVIRONMENTAL
CHANGE, AND THE GREAT OAKLAND
HILLS FIRE

Gregory L. Simon

UNIVERSITY OF CALIFORNIA PRESS

University of California Press, one of the most distinguished university presses in the United States, enriches lives around the world by advancing scholarship in the humanities, social sciences, and natural sciences. Its activities are supported by the UC Press Foundation and by philanthropic contributions from individuals and institutions. For more information, visit www.ucpress.edu.

University of California Press
Oakland, California

© 2017 by The Regents of the University of California

Library of Congress Cataloging-in-Publication Data

Cataloguing-in-Publication data on file at the Library of Congress.

ISBN 978-0-520-29280-2 (cloth ; alk. paper)
ISBN 978-0-520-29279-6 (paper ; alk. paper)
ISBN 978-0-520-96616-1 (ebook)

Manufactured in the United States of America

25 24 23 22 21 20 19 18 17
10 9 8 7 6 5 4 3 2 1

For Dimitri and Gabriel

In honor of all who experienced loss and trauma
from the Oakland Hills Fire

Contents

	Preface	ix
	Acknowledgments	xi
	Introduction	1
PART I	FLAME AND FORTUNE IN THE AMERICAN WEST: AN INTRODUCTION TO THE INCENDIARY	
1.	The 1991 Tunnel Fire: The Case for an Affluence-Vulnerability Interface	11
2.	The Changing American West: From "Flammable Landscape" to the "Incendiary"	39
PART II	ILLUMINATING THE AFFLUENCE-VULNERABILITY INTERFACE IN THE TUNNEL FIRE AREA	
3.	Trailblazing: Producing Landscapes, Extracting Profits, Inserting Risk	55
4.	Setting the Stage for Disaster: Revenue Maximization, Wealth Protection, and Its Discontents	71
5.	Who's Vulnerable? The Politics of Identifying, Experiencing, and Reducing Risk	89

PART III	HOW THE WEST WAS SPUN: DEPOLITICIZING THE ROOT CAUSES OF WILDFIRE HAZARDS	
6.	Smoke Screen: When Explaining Wildfires Conceals the Incendiary	109
7.	Debates of Distraction: Our Inability to See the Incendiary for the Spark	130
PART IV	AFTER THE FIRE: THE CONCOMITANT EXPANSION OF AFFLUENCE AND RISK	
8.	Dispatches from the Field: Win–Win Outcomes and the Limits of Post-Wildfire Mitigation	149
9.	Out of the Ashes: The Rise of Disaster Capitalism and Financial Opportunism	168
	Conclusion: From Excavating to Treating the Incendiary	185
	Notes	209
	References	227
	Index	239

Preface

As I write this, nearly 150,000 acres of California are burning. Over ten thousand firefighters are responding to twenty-one active fires and counting. More than thirteen thousand people are under evacuation advisement and over six thousand structures are currently threatened. These numbers seem to rise with each news cycle. Governor Jerry Brown has just issued a state of emergency and is mobilizing the National Guard to assist in the statewide disaster response. The largest of these fires at the time of writing is the Rocky Fire, a 65,000-plus acre blaze (over 100 square miles) burning in the Lower Lake area north of San Francisco. One firefighter, a thirty-eight-year-old father of two, has died battling the Frog Fire farther north near the Oregon border.

According to Governor Brown, California's acute and persistent drought conditions have "turned much of the state into a tinderbox." Indeed the flammable nature of California could not be clearer than it is from my current vacation vantage point looking across Lake Tahoe near the California/Nevada border. The lake's shoreline is separated from the soaring ridgeline and blue sky by a thick strip of grayish smoke. This hazy dividing line between water and sky is a by-product of several fires burning in nearby "tinderbox" environments of the Sierra Nevada foothills and higher-elevation forests. But while these fires rage on and obscure my view of Lake Tahoe, I quickly remember that California and the American West have always had fires. Episodes of drought are not uncommon. And even during times of normal precipitation, fires periodically sweep through landscapes—sometimes burning thousands of acres at a time. In this way much the West has been (and will for the foreseeable future be) a so-called tinderbox.

Sitting at the lake's edge I am reminded of what makes *this* fire season—like each preceding year—seem so unique and urgent, particularly given the region's historically active fire regime. The answer is simple: people like me, and the millions of others who visit, live, own, build, plan, develop, and market property in traditionally fire-prone areas like Lake Tahoe. These groups and individuals are the driving force turning historically recurring fires into devastating fire *disasters*. Our policies, planning decisions, and cultural preferences generate these fire risks and the massive human and economic costs that accompany them.

Although the prevailing narrative on fire is that flammable (or "tinderbox") environments are somehow produced by forces greater than ourselves—a by-product of years of drought, for example—the truth is that suburban and exurban homes, residents, and the development forces behind them are not merely victims of the unassailable forces of nature. Rather, humans assist in the creation of environments where fires become seen as negative, detrimental, and disastrous. It is because of human activity that landscapes historically subject to fire are now viewed as a victimized tinderbox. To be sure, this is a contradictory position, as we are creating the conditions we actively fear, resist, and attempt to mitigate. It is true that climate change and persistent drought are currently making matters worse. But a review of media reports, such as those covering the current fire season in the West, makes clear that society's ravenous appetite to develop these historically high-risk areas is frequently let off the hook and exonerated from responsibility. In the pages ahead I argue that this is a disingenuous, costly, and dangerous game to play.

Acknowledgments

This book would not have been possible without the help of countless individuals, too many to list here, who influenced the project in important and profound ways. My parents and sister gave incredible amounts of support over many years and decades, all of which instilled in me a sense of commitment, confidence, and curiosity to see this project to completion. I am grateful to Dylan for her steadfast encouragement and support throughout the project, including travel periods and writing days both long and short.

This project would not have come to fruition without members of the Tunnel Fire community as well as other fire victims in the region who took time to peel back and revisit often difficult memories in order to shed light on life before, during, and after hazardous fire events—Jesse and the entire Grant family in particular, whose original home still lives vividly in my memory. Thanks to Vicky for taking the lead with the adorable barbarians throughout this period. Many local and state fire service members must be thanked for providing key and provocative insights.

I acknowledge Christine Erikson for her companionship, good humor, and immensely refined and adaptive interviewing chops while in the field, particularly for material that appears in Chapter 5; Peter Alagona, Matthew

Booker, and Robert Wilson for their insights during early stages of this project at Stanford's Bill Lane Center for the American West; and Peter too for sharing many miles of trail-inspired conversation and smoky scenic overlooks in the Sierra Nevada. Thanks also to Richard White and David Kennedy for their support during this time, and Richard too for continued support over nearly a decade, including introducing me to the Spatial History Project at Stanford University. Thanks to Zephyr Frank and others with the Spatial History Project for their support of the Vulnerability-in-Production project. Many students supported this research and most of the heavy geospatial lifting. This project truly would not be where it is today without the impressive and inspiring work of CU Denver and Stanford students. These include Kathy Harris, Allie Hausladen, Tyler Kilgore, Emily Kizzia, Eric Ross, Alejandra Uribe, Melissa Wiggins, and others.

The staffs at the Bancroft Library at UC Berkeley and the Oakland Library History Room were extremely helpful throughout this project. Several members of the geography departments at UCLA and UC Berkeley provided incisive feedback, which pushed my thinking about the affluence-vulnerability interface. Thanks also to Stephanie Pincetl and others at UCLA's Institute of the Environment and Sustainability for their generosity over the last several years. The Yale School of Forestry and Environmental Studies community also offered helpful comments during this time. Sarah Dooling made invaluable and challenging comments that pushed this project intellectually in new and exciting directions. So too did Tim Collins and many others who contributed to the book *Cities, Nature, Development: The Politics and Production of Urban Vulnerabilities*. I owe considerable thanks to the entire faculty in the Geography and Environmental Sciences Department at CU Denver for providing the resources, work environment, and friendships required to complete this project. I am grateful to Julie Guthman, Jake Kosek, and Rebecca Lave for including this book in the *Critical Environments* book series. Last but certainly not least, I want to thank Merrik Bush-Pirkle and the entire staff and editorial group at UC Press for skillfully guiding this project through to completion.

Introduction

Early afternoon has arrived in a swirl of charcoal skies and brown haze. Wind-whipped orange embers soar hundreds of feet in the air. Most are lost from sight, enveloped and extinguished by the suffocating column of smoke billowing and spilling over the ridgeline directly in front of me. Other glowing embers drop from the sky out of view into a canyon or atop a ridgeline, the fate of these sparks forever lost to the fire. Out of the sky fluttering debris—black and gray wisps that appear like burned bits of newspaper—drizzle steadily downward. Like the first signs of winter snow these softly falling objects portend a larger storm on its heels: a soothing aesthetic belying its violent source searing just over the hilly crest to my north.

I am standing alone in a large five-way intersection one block uphill from my childhood home—a 1940s, two-story, wood-shingled home nestled into one of the many densely vegetated canyons populating this coastal hill range in Oakland, California. A large eucalyptus tree twists upward from the middle of the intersection. It is a tree I have seen a thousand times before. But today it looks different. The single, thick-trunked eucalypt typically looms like a sentient guardian over the neighborhood; at this moment, in the eerie, grayish-yellow glow, the tree looks isolated, helpless, and unsuspecting. A sudden gust rustles leaves and bark litter at its base, sweeping them onto and then off its golden escarpment as if to nudge the eucalypt—a subtle warning of an impending tempest.

As I quickly turn and jog back downhill, my breath runs deeper, not because of the physical activity but because of a building sense of anxiety and fear. I now realize it is not the eucalypt that feels isolated and helpless; those scattered, wind-blown leaves and bark fragments have stirred up my own unease. Moving quickly now, with each breath I smell the pungent and heavy scent of a chaotic and disorderly fire, a calamitous concoction of burnt, woody, and sickly sweet odors that could only be produced by a wildfire burning indiscriminately through backyards, bedrooms, and basements. The smell is at once all wrong and all too real. My family is out of town and a distressing weight of responsibility and panic sets in. As I arrive at the house, my anxiety deepens. Will my family's home, our belongings, our cat—will I myself—contribute to this terrifying smell and wind up falling gently from the sky like ominous snowflakes onto other nervous, clambering residents?

The increasingly chaotic environment outside fades away as I enter my home and close the front door behind me. Misty is sleeping on the couch, her nose tucked under her front paw; she is sweetly oblivious to the calamity unfolding outside. I sit for a moment, gently petting her gray coat, and ponder how I came to be in this wildly unfamiliar and improbable predicament. Earlier that morning I had driven two friends home from my parents' house. I traveled north to drop off a friend near the Oakland/Berkeley border and then proceeded to my second friend's house in North Berkeley. Returning home through the UC Berkeley campus toward Oakland, I noticed a subtle dimming of the sun overhead. This was a strange sight as the sky was clear and cloud-free when I departed the house an hour earlier; it had been the kind of warm, sunny morning that is typical during late fall in the San Francisco Bay Area. This sudden reduction in solar radiation was startling, as was the hazy, attenuated nature of the sunlight hitting my windshield. Turning east toward the East Bay hills, I noticed a large column of smoke rising toward the sky. Unaware of the source and, more importantly, the seriousness of the situation, I continued to drive home.

Now sitting on the couch—the winds outside whipping through the redwood tree—I realize that the hazy sunlight on my windshield had been a dystopian exhortation from the hills above, an ominous sign reaffirmed by the smoky view from the solitary eucalypt on the hilltop. Though still unsure about the source or full extent of the storm, I begin to contemplate my next move. Just then a blaring police siren and bullhorn ring through the window behind me. Turning to look I can see a policeman standing outside of his vehicle directly in front of the house, his uniform whipped by unyielding winds careening downhill from the fire. He is holding on to his sunglasses and yells into the bullhorn, "If you are still in your home you must evacuate *now!*" I jump off the couch and head outside through the backdoor.

Walking into my backyard I scan above for signs of danger, as the sky overhead seems the best and most reliable indicator of my precariousness. Beyond a row of pines I am able to look north toward the source of the fire. To my relief I do not detect any flying embers in the wind-whipped mixture of haze and smoke. From one or two blocks up the road I hear the same police warning, along with some voices coming from behind my backyard gate. These sounds are followed by slamming doors and a car engine, which quickly fades away. More voices. Then more car doors. Then the sound of yet another car backing out of its driveway and speeding out of the neighborhood. I pass through the back gate toward these sounds and am elated to see my best friend, Jesse, across the street emerging from his own home. We share a brief adrenaline-filled conversation, which shifts quickly from "Can you believe this?" to "We need to do something!"

Overhead a high layer of smoke continues to gather and circulate. As we stand in the middle of the street between our two homes, several pieces of white debris, similar to what I had witnessed up the hill only minutes earlier, now fall from the sky. It appears the fire and its swirling detritus are getting closer. Suddenly an elderly lady approaches, waving her arms. Clearly in a panic she describes her inability to get her infirm husband out of their house. We follow her home where we pass a car in the driveway. The backseat is packed full—a number of unclosed boxes bulge and protrude from underneath a neatly placed checkered blanket. The passenger door is open but the passenger seat remains empty. Down the steps, in the front doorway, her husband is sitting in a wheelchair, helpless and unable to ascend the roughly twenty stairs leading to the car. Jesse and I run down to his side. He is frail in clothes that no longer fit his gaunt frame. Sliding our arms under his lap from either side, we clasp hands and lift his body, with surprising ease, up and out of the chair. He gently places his arms around our shoulders but is too weak to hold them there. They slide down our backs and dangle at his sides. Leaning together to hold him in place, Jesse and I hoist him up the steps toward the open car door. We gently maneuver him into the front seat. As we walk back, we hear the now familiar sounds behind us: doors slamming, a car starting, and an engine fading quickly up the road.

Despite the menacing warning signs we decide to walk past our homes and up the hill toward the source of the fire. We make it one hundred feet or so when we are stopped in our tracks by a disturbing view. Whereas the intersection at the top of the hill had previously revealed a high blanket of swirling smoke and gently falling debris, this same location is now enshrouded in a much darker mass of gray. As we peer one block up the road through the smoky clouds, an even more troubling site emerges: faint wisps of bright, flickering orange stretch, bend, and retract with the swirling winds. No longer just a vantage point for viewing the source of the smoke, our neighbors' homes have now *become* the source of smoke.

We head back down the hill and are confronted by Jesse's father, who proclaims urgently that embers are now dropping in their backyard. As Jesse and his dad head up their driveway, I overhear comments about checking for water pressure and spraying down the property. I figure I should do the same and head across the street to my house where I pick up a hose neatly coiled next to the garage. I turn the faucet and wait. A few gurgles later a small pool of water gathers slowly at the nozzle of the hose and pours gently over its edge. I check the line for knots but don't find any, so I turn the hose on all the way. The water bubbles again, this time spilling over the top. I press my thumb over the nozzle to coax more water pressure but this only amounts to a lazy one-to-two-foot stream of water—not nearly enough to reach the wood shingle siding above me, let alone the wood shake rooftop thirty feet overhead.

I abort the watering mission and start recoiling the hose. Debris is now flying downhill away from the fire, picking up speed and whipping past structures and swaying branches. An occasional red ember drops and weaves its way through the airborne debris, landing and bouncing erratically along the road. As I finish with the hose, the same patrolling police officer races up the road. He gets out of the car with an incredulous and angry look. He tells me to leave immediately. Do not pack anything. Just get my keys and go. I ask if I can run inside and get my cat. Growing irritated, he tells me to hurry and says he is going to wait until I come out of the house before leaving. I race inside to grab my keys, scoop up my cat, and put her in the cat box. I do not take anything else.

Cat in hand, I turn one more time toward the house. I peer up at the redwood, the coast live oak, and the pittosporum, all standing guard over my family's home. It is as though I am seeing this scene as well as recording it. I am overcome with a sudden sense of resignation: this will be the last time I have this view. I imprint on the scene for a few more seconds and then get in the car. Slamming the door, I start the engine and drive hastily away.

That evening, removed from my front-row seat, I follow the news on television from my grandmother's house across the bay in San Francisco. Sitting together in her kitchen we flip between TV news channels covering the now massive blaze. At one point we watch a familiar-looking structure engulfed in flames. A news crew must have moved behind the safety perimeter to record this small piece of the devastation. We recognize the shape and front entrance of the house before the news reporter confirms our suspicions and describes his crew's position on our street; the burning home in question sits directly across the street and to the south of my family's home. On the other side—directly to the north—is Jesse's house, whose backyard was catching fire when I left. I decide to call his home number, thinking that might reveal a clue about the home's fate. Sure enough, instead of getting a ring, busy signal, or answering machine, I get an empty dial tone and an abrupt click. As much as my heart wants to believe otherwise, it appears that Jesse's home with its wood shingle exterior and solid redwood interior suffered a similar fate. Given the position of our home between these two burning properties, we are all quite certain of a similarly ruinous outcome.

Arriving at this sobering conclusion, we head to the airport to pick up my parents who are arriving from out of town. After several minutes of anxious waiting at the gate, we finally greet them as they exit the jet-way. Expecting a curbside pickup they seem rather confused by our presence at the gate. My grandmother calmly tells them to sit down. And with all the subtlety of a sixteen-year-old I blurt out, "Mom, Dad. Umm . . . I'm pretty sure our house burned down." They receive this information as well as can be expected—disbelief, a profound sense of sadness, relief for my safety, and a lingering hope that perhaps our calculations are wrong. Needless to say, although we all go to bed that evening, nobody is sleeping. I for one lie wide awake, fixating on how I failed to take a single item—save the cat—from the house. No pictures. No keepsakes. No passports or emergency files. It feels as though I just failed a crucial life test.

Assuming the worst, the next day I sneak through established disaster perimeters to see the full extent of the fire's damage. As I emerge over a ridgeline, the harsh and startling reality of the previous day's devastation unfolds in front of me (see Figure 1 for partial aerial view of destruction). Protruding chimneys, stripped bare and exposed to their grayish foundation, like tombstones in a cemetery, cover the landscape. Many trees are singed to the branch and trunk. Nearly every home is gone. The fire had moved in capricious ways, however, sparing arbitrary structures along the way. From my vantage point atop the ridge I am able to peer into a small, intact basement closet—the remainder of the house is lost. Through the closet's open door a stark white dress shirt hangs unbothered against the pallid background.

Looking farther south, I scan cinereous slopes and valleys for the ruins of my family home. It is difficult to pinpoint exactly where I am looking, as nearly every identifying landmark has been wiped off the now collapsed and scorched landscape (see Figure 2). Growing more disoriented, my gaze locks on to a familiar-looking structure below. Can it really be? Is that . . . what I think it is? To my shock the home I left hastily a day earlier remains fully intact alongside a neighbor to each side. Inexplicably the surrounding blaze never ignited the home's wood-shingled rooftop and sidings. This unfathomable discovery is too good to be true; I am overjoyed to see the structure and all our possessions in one piece. This euphoria is quickly lost to the sight of smoldering ruins all around, including our neighbors' and friends' homes forever lost to the now extinguished wildfire. Red embers falling into my friend's backyard had indeed spread to his home. And the news footage of our neighbor's house ablaze on television did not lie. The surrounding damage is so extensive, in both severity and scope, that for a moment I do not know which direction to leave the neighborhood. I later learned that the first friend I had dropped off the previous day was evacuated from his home within an hour due to fast-moving flames. A couple of hours later his house had burned to the ground along with all of the neighboring homes.

Figure 1 (top). Looking east: an aerial panorama captures a portion of the Tunnel Fire area. The blaze began near the upper left-hand corner of the image and stretched out of view in all directions.

Figure 2 (bottom). Walking through the burn area revealed a neighborhood reduced to a cinereous state, comprising charred vehicles, debris, vegetation, and chimneys.

In retrospect, at the time of the fire the source of the pyro-precipitation and acrid aroma was unclear. While I recognized the situation as serious, I did not (indeed, could not) appreciate that behind the churning and fast-encroaching sallow mass of smoke lay a savage storm, with thousands of cherished homes and memories left smoldering in its wake. Beneath this foreboding curtain of smoke over two dozen lives were being lost.

Yet given the day's conditions and the previous century of extensive residential development, the outcome of the 1991 Tunnel Fire was both unsurprising and in many ways inevitable. This is the nature of much of California and the American West: *we can see fires without seeing the fires.* They are coming. They will continue to come. Our past and ongoing land-use planning decisions—premised on profitable suburban and exurban land development—will see to that.

On that fateful Sunday afternoon a thick charcoal veil concealed my view of the Tunnel Fire. In much the same way a tenacious storm of development and increased flammability rages behind a curtain of ongoing political, cultural, and scientific debates about wildfires and their causes and consequences. Hidden behind these debates are persistent suburban growth trends and financial incentives that undergird an increase in fire risk and mitigation costs around the region. The chapters ahead constitute a political ecology of fire, bringing these processes into sharper relief. The hope is that they will spark a renewed engagement with proactive management approaches that directly confront the systemic causes of fire risk.

PART I Flame and Fortune
in the American West

AN INTRODUCTION TO THE INCENDIARY

1 The 1991 Tunnel Fire

THE CASE FOR AN AFFLUENCE-VULNERABILITY INTERFACE

A FIRESTORM FOR THE AGES: THE OAKLAND HILLS (TUNNEL) FIRE

The 1991 Oakland Hills Firestorm (henceforth Tunnel Fire) is perhaps the most significant urban wildfire in United States history. Located in northeastern Oakland, California, and stretching northward into the city of Berkeley and east into neighboring Contra Costa County, the Tunnel Fire destroyed more than three thousand dwelling units and killed twenty-five people over a twenty-four-hour period (see Map 1 and Map 2). The fire's origins were innocuous enough—a small brushfire that had burned and been suppressed the previous day. The response to this small outbreak had been extensive and decisive. Aware that flare-ups could occur throughout the night, city fire crews were on high alert with water lines in place.

A change in firefighter shifts early on October 20 forever changed history. Throughout the evening and into the morning the previous day's fire was stirring and smoldering beneath the singed surface. As new engines moved back into the hills to treat the burn area, hot spots could be seen igniting into flames all around. More support units were immediately

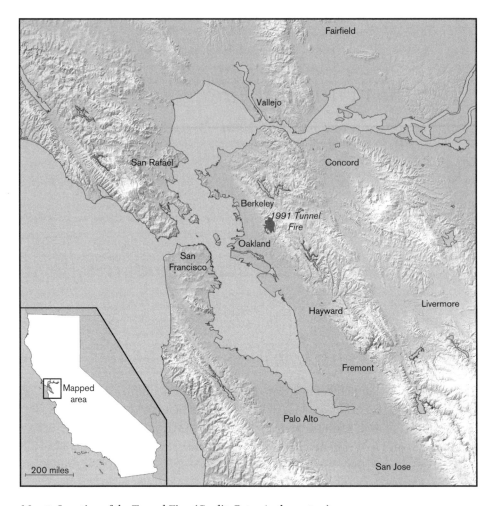

Map 1. Location of the Tunnel Fire. (Credit: Peter Anthamatten)

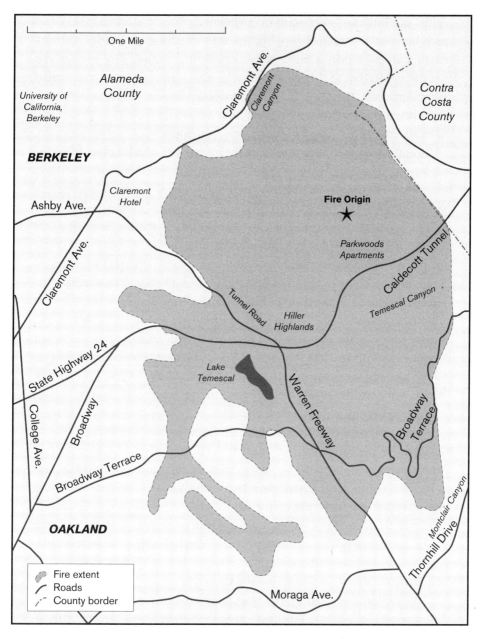

Map 2. The Tunnel Fire area extent, major land features, and fire origin. (Credit: Peter Anthamatten)

called up, but they were soon overwhelmed. Strong winds happened to pick up that morning, whipping through the hillside, quickly turning small spot fires into a mass of fast-spreading flames. The fire ignited nearby vegetation and began its steady march into adjacent residential neighborhoods. The fire was both relentless and voracious, burning nearly eight hundred structures in the first hour alone and more than three hundred structures per hour over the next seven hours.[1]

Twenty-five years later, the Tunnel Fire has left a lasting legacy in the region as the largest wildfire—in terms of numbers of dwellings destroyed—in California's history. To be sure, this is a dubious distinction given California's long record of frequent and intense firestorms.

The superlatives go on. The 1991 Tunnel Fire is the single most costly *wildland* fire in modern U.S. history.[2] In adjusted 2012 dollars the fire is estimated to have generated $2.5 billion in losses. The next four largest wildland fire losses range from $1.2 billion to $2 billion.[3] To compare the relative intensity of the Tunnel Fire with these other events consider the following statistic: the Tunnel Fire burned just over 1,500 acres (2.3 square miles) while the next four most costly wildland fire events ranged between 48,000 and 375,000 acres.[4] With well over three thousand structures burned (the most in U.S. wildland fire history) in just 1,500 acres, the severity and intensely "social" and "urban" nature of the fire is both unmistakable and unparalleled.

Across the region, nation, and even internationally, the Tunnel Fire (or "Oakland Hills Firestorm" or "East Bay Hills Firestorm" depending on who is reporting) remains *the* urban wildfire reference point in U.S. history. While other large events such as the Chicago and San Francisco fires are infamous for their scope, intensity, and devastation, they are considered strictly urban fires—not wildfires, which connotes a link to adjoining "wild" open space. The notoriety of the Tunnel Fire as a "worst-case scenario" wildfire sparked substantial changes to firefighting protocols in departments and agencies across the United States.

The Tunnel Fire's significance has much to do with its geography. The fire area sits in a precarious location at the hilly edge of the core metropolitan cities of Oakland and Berkeley, along the eastern side of the San Francisco Bay. Much will be made of historical land use planning and residential development activities in these areas. However, three main characteristics should be highlighted up front. First, this is a *wildland-urban interface* (WUI) area, meaning it resides at the interface between dedicated urban developments and managed open space. Residential communities sit within and astride a landscape comprising dense forest cover, woodlands, grasslands, and coastal scrub. These housing and wildland landscapes have for many decades spread into each other, intermingling to produce an environment notable for its high fuel load. The area is also remarkable for its hilly terrain and steep canyons, which have the dual effect of aiding fire spread while also hindering efforts to prevent and respond to fire.

Second, it is a centrally located urban area, meaning it is not a new subdivision at the periphery of the San Francisco Bay Area but rather a century-old (and in some locations older) neighborhood containing high-density residential housing and narrow roads that are characteristic of older, core urban areas. (Of course, at the time of its development it was a suburban enclave and so fits within historical and ongoing suburban leap-frog sprawl patterns.) Beyond the adjacent swath of regional parkland and state responsibility "wildland" areas lies even more urban, suburban, and exurban development. While this area was once considered a suburban respite from the San Francisco grit and grind (and it does still retain those characteristics), the suburban fringe has expanded outward for many miles, cities, and even counties in all directions.

Third, the region sits within a Mediterranean climate that contains long durations without precipitation. From 1981 to 2010, total rainfall over the four-month period of June 1–September 30 averaged 0.39 inches, most of which typically precipitates from fog. Thus by the time October arrives each year, the area is usually water deprived and composed of dry tree cover, grasses, shrubs, and ground litter. Making matters worse are periodic Diablo winds, which bring a dry, easterly airflow from the hot Central Valley through the San Francisco Bay Area. These winds—sometimes at speeds of up to 60 mph—increase fire risk by further desiccating the land environment and significantly increasing the spread potential of small spot fires. These are the winds that reignited remnant embers and turned the previous day's seemingly innocuous ground fire into the destructive and record-breaking Tunnel Fire. When viewing these physical attributes in aggregate, we see that geography matters. The

Tunnel Fire area retains a level of home density indicative of core urban neighborhoods (i.e., greater home exposure) while also having close proximity to dense and flammable tree cover that is suggestive of settlements at the urban fringe.

But it is the region's history that makes the Tunnel Fire particularly compelling as a subject of analysis. Dating back to 1900, the area's fire regime indicates a high frequency of wildfires in the Oakland Hills region. Between 1900 and 1991, for example, twelve fires were recorded in close proximity to the area consumed by the Tunnel Fire. This includes major fires in 1923 and 1970 that swept through much of the same area eventually seared by the Tunnel Fire. Some homes burned and rebuilt during earlier fires were again destroyed in the 1991 blaze. This long history of persistent conflagrations raises serious questions about urban development, land use planning, and resource management efforts that have significantly increased fire risk by placing tens of thousands of homes in already fire-prone landscapes (see Map 3). This is a development trend that produces social risks while simultaneously increasing the costs of reducing social vulnerability. Indeed it is a profoundly contradictory positive feedback loop—intentionally producing an undesirable and costly outcome. But it is also a process that holds an internal logic driven by a set of underlying incentive-based economic structures. It is precisely this structural logic that this text intends to excavate and highlight.

The Tunnel Fire presents an opportunity to thread a consistent narrative and distinct reference point throughout the book. The Tunnel Fire is also a unique case and the histories, processes, events, and lessons that emerge from the Oakland and Berkeley hill areas do not translate neatly to other regions of California and the West. It is therefore important not to overstate the explanatory power of this single event when evaluating issues such as the causes and implications of fire risk across the entire region. (For example, over three thousand units destroyed in only 1,500 acres is simply unmatched in modern U.S. history.) On the other hand, what happens in Oakland does not stay in Oakland. As we will see, social risks, economic opportunism, fire mitigation policies, scientific debates, and basic fire behavior all (in most instances) transcend political and environmental boundaries. The Tunnel Fire therefore provides a useful starting point for analysis and a critical source of information from which larger lessons,

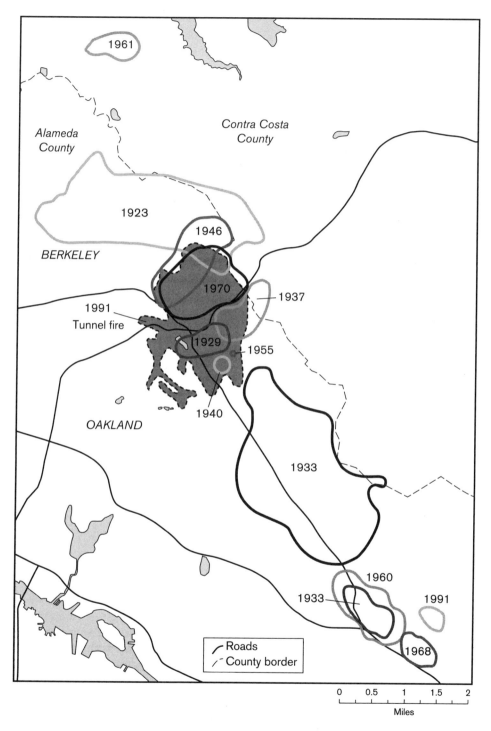

Map 3. Periodic fires in the East Bay hills over the past century reveal the persistent threat of conflagrations. (Credit: Peter Anthamatten)

perspectives, and theories can be developed. It also provides a geographical mooring for an otherwise expansive, complex, and seemingly unwieldy set of social-ecological entanglements. The Tunnel Fire is thus leveraged as a means of engaging with in-depth historical and contemporary analysis—grounded in place—that prevents the text from producing an overly synoptic, superficial, or fragmentary analysis. Throughout, the book zooms in and out, shifting between close readings of events and landscapes in the Tunnel Fire area and important developments and trends in other regions of the West. This multigeographical approach (which admittedly retains a strong emphasis on the Tunnel Fire) enables the productive conjoining of idiographic, descriptive, and place-based analysis with nomothetic, inductive, and generalized interpretations.

URBAN DEVELOPMENT, NATURE, AND WILDFIRE:
THREE CRITICAL INSIGHTS

Fire at the urban periphery is a complicated matter, and understanding its dynamic and complex social and environmental underpinnings is certainly no easy task. Efforts to fully comprehend how we come to know and respond to the threat of fire are perhaps even more challenging. Given these complexities, *Flame and Fortune* offers three principal insights about the relationship between fire, society, and environment at the city's edge. By focusing on these critical insights, the book will in turn contribute to a single primary analytic objective: the need to augment assessments of the wildland-urban interface with dedicated analysis of the *affluence-vulnerability interface* (AVI; see below) and *the Incendiary* (Chapter 3). Let us look briefly now at these three insights in turn.

Lucrative Landscapes at the Urban Periphery:
Taking Profits, Adding Risk

Suburban landscapes are lucrative landscapes. They are areas that generate high levels of profit, revenue, and wealth for interested parties near and far. From early land use extraction activities to contemporary private fire mitigation services, diverse groups extract profits from these regions, thus leveraging the suburban landscape as a source of prosperity and increased affluence. These profit- and revenue-generating activities are certainly not benign. Over time the generation of financial benefits has coincided with the production and maintenance of risk and vulnerability. This is the nature of urban growth under capitalism—it produces both beneficiaries and discontented groups, simultaneously. Still further, in many instances we see that one depends on the other: efforts to increase affluence oftentimes necessitate elevating levels of fire risk, and higher risk and vulnerability levels frequently spur opportunities to generate further financial gains.

Factors influencing increased social vulnerability and higher fire risk mitigation costs are inextricably tied to ever-changing profit-seeking practices and diverse forms of economic opportunism. This is an important foundational insight because it provides analytic space to reveal the systemic causes of social vulnerability to fire, as well as the policies, practices, and perspectives that enable the development of these risky urban landscapes. The 2016 Fort McMurray fire in Alberta, Canada, is a recent example of a devastating wildfire impacting a region initially developed for its ability to generate profits for government and corporate interests alike. The area of Fort McMurray, which lies several hundred miles north of Calgary, grew rapidly in population and size over the past several decades to support large-scale extraction of oil from an enormous subterranean tar sands deposit. Like so many fires in the West, the pursuit of resources and the conversion of landscapes were fundamental drivers of this "hazardous" event.

Moreover understanding the root causes of fire risk is an important first step toward substantively reducing future costs associated with patterns of material accumulation and seemingly unfettered urban expansion—as the old adage goes, "You have to understand the problem before you can find the solution." Directly confronting these past and ongoing processes will require a rhetorical and tactical shift. It will entail ceasing to treat portions of the American West as flammable landscapes (resulting in management frameworks that merely treat development *symptoms* at the WUI) and instead shifting to the treatment of the West as the "Incendiary"—a regional depiction emphasizing the area's history, foundational characteristics, and underlying socioeconomic drivers that produce elevated flammability; that is, viewing the landscape and the forces behind its transformation like an arsonist that must be directly undercut, intensively treated, and *not* merely adapted to.

The Persistent De- and Repoliticization of Fire and Its Production

Contemporary management and scientific discourses on fire *depoliticize* the Incendiary and the political economic root causes of fire disasters. Depoliticization refers to the process of stripping an issue or event of a core political or controversial underpinning. This allows particular foundational explanations of social-environmental change—in this context, processes related to the affluence-vulnerability interface and its associated controversies—to go unnoticed and unchallenged. In the case of wildfire we oftentimes get distracted from key debates concerning the social origins of fire risk by other management and science controversies. We can understand these corollary or alter-debates as contributing to a process of *repoliticization,* as a series of other contested issues come to occupy the discursive arena of disagreement and dispute. This repoliticization largely obviates substantive discussions of broader structural issues and drivers. Concealing the role of private material accumulation and urban growth regimes, for example, ultimately has the effect of rendering wildland-urban interface fires as simply the natural order of things. We are left tinkering around the edge of the problem, constantly putting out little fires instead of grappling with the root cause of the major blaze itself.

This analysis complements other studies that have examined the politics of managing natural resources, landscapes, and disasters. Environmental debates, for example, are often portrayed as proxy debates for broader disagreements over the proper role of the government and free market, or the importance of individual freedoms and private property rights.[5] Arguments over the fate of individual species, for example, serve as convenient and tractable sites for engaging in, and ostensibly "settling," these broader debates. This text adds to this discussion by suggesting that not only are broad debates fought in small arenas, but in fact the acrimony found in these small arenas can distract us from addressing larger disagreements, tensions, and contradictions. These alter-debates may actually *prevent* us from reconciling (and thus directly confronting) larger issues concerning the political economic and social origins of fire risk.

Clarifying Vulnerability as a Dynamic, Variegated, and Malleable Concept

Vulnerability is typically treated as a static condition and applied to households, communities, and cities in ways that oversimplify the complex manner in which it is produced and experienced over time and space. In *Flame and Fortune* vulnerability is shown to exist as both a process and a condition. Expanding the ontology of vulnerability to include both these aspects is important because it allows us to understand the complex relationship between different vulnerabilities as they unfold and interact spatially and temporally. First, vulnerability is presented as a recursive and relational process—embedded within regional environmental and development histories—that is always in production, at play, and inscribed unevenly over time and space. Vulnerability is thus much more than simply an effect of planning, produced outcome, or material inscription. Factors causing increased vulnerability in one area may deepen (or maintain, or reduce) risks for individuals in another area or in subsequent time periods. For example, socioeconomic conditions leading to increased risks in hill areas of Oakland also contribute to elevated risks in the city's flatland. Viewing different manifestations of vulnerability as merely isolated conditions would foster a truncated and thus incomplete account of important geographical and historical events and connections.

Second, vulnerability is variegated across space and is manifest in unique ways within diverse households. Differentiating vulnerabilities from one home to the next is important because it prevents the tendency by many to overgeneralize (or simply ignore analytically) entire populations and their experiences with risk. This is particularly true in suburban areas of the West that appear to be affluent and thus somewhat immune to any substantive or acute form of vulnerability. Understanding that conditions of vulnerability are distributed (and felt) unevenly between households renders such statements both erroneous and simplistic. In the Tunnel Fire area, for example, the concept of variegated vulnerability signifies that neighborhoods contain diverse households, each with distinct biophysical settings, family histories, physical and emotional fragilities, and financial resilience. This household-level heterogeneity winds up influencing how homes and individual residents experience and respond to the fire and its aftermath.

SUPPLEMENTING THE WILDLAND-URBAN INTERFACE WITH THE AFFLUENCE-VULNERABILITY INTERFACE

Together these three critical insights contribute to a single, larger analytic objective: to explicate the benefits of an affluence-vulnerability interface (AVI) approach to the study of society, nature, and wildfire. Currently three capital letters, W-U-I, dominate public conversations concerning the transformation of urban peripheries and the resulting increases in residential exposure to (and costs of) wildfires. The term *wildland-urban interface* is one the most ubiquitous phrases circulating through the suburban and exurban wildfire management discourse. It is *the* land designation used to connote the uneasy overlap of human settlements (composed primarily of residential infrastructure) with traditionally undeveloped or wild (and often already fire-prone) environments.

The wildland-urban interface is a rather recent concept and geographic construct and is described by the National Wildfire Coordinating Group as "the zone of transition between unoccupied land and human development. It is the line, area, or zone where structures and other human development meet or intermingle with undeveloped wildland or vegetative fuels."[6] In fact the term was only formalized in federal policy as a distinct area of interest in 1987—just four years prior to the Tunnel Fire. When the U.S. Department of Agriculture presented its budget to Congress in February of that year, it announced six new research initiatives. The second of those initiatives was titled the "Wildland-Urban Interface." This area of emphasis called for increased public attention to areas where urban settlements intermixing with wildlands contain "major problems in fire protection, land use planning, and recreation impacts."[7]

The WUI is actually one of two primary land designations at the urban fringe. The other is the wildland-urban *intermix*. The *interface* is characterized by greater home densities, with structures located adjacent or in close proximity to one another. As a collective unit these neighborhoods press up against large, undeveloped open spaces. The *intermix* connotes an area with greater distance between homes, where the home structures themselves are embedded within surrounding wildlands.[8] While the interface landscape is dominated by homes, the intermix is composed primarily of open or forested landscapes. Despite their differences both the intermix and interface are conflated within a single WUI land designation (see Figures 3 and 4).[9] A precise and singular definition of what constitutes an interface or intermix area remains elusive, and different studies have used varying housing density thresholds to determine such designations. Whatever land attributes are used, in the context of fire it is generally understood that areas within a half mile of qualifying landscapes are still considered part of the WUI as they are susceptible to the impacts of swirling smoke and flying brands.

The establishment of a WUI land designation—despite its somewhat malleable definition—is certainly not without policy consequence. This designation has provided an easy-to-map and thus legible geographical area supporting the structured implementation of a number of land use and forest management practices. These include early efforts to extend the U.S. Forest Service's "fire exclusion paradigm" into developed areas through dedicated fire suppression–based home protection[10] and more recent "Fire Adapted Communities" approaches premised on providing activities that increase community education, preparedness, and resilience to periodic fire events.[11] The WUI has become a scrutinized landscape in part because it challenges conventional and binary nonurban/urban and public/private land classifications. In an effort to reduce the opacity, disorder, and novelty contained within these landscapes and in order to achieve greater analytic and managerial clarity, the WUI has emerged as a useful land classification—a conceptual container within which we can study, interpret, and manage the messy and complex transition from nonurban to urban and public to private.

A Political Ecology of Fire

A shift in perspective is in order. *Flame and Fortune* argues for a move away from the wildland-urban interface as *the* principal organizing

Figures 3 and 4. An all-too-familiar scene. Wildfires threaten homes in wildland-urban intermix areas (top) and wildland-urban interface areas (bottom) of the American West. Both settings fall under the broader category of the WUI, or wildland-urban interface.

framework guiding the management of wildfires at the urban periphery. Instead the following chapters suggest the adoption of an affluence-vulnerability interface approach. This political ecological shift promotes a conceptual shift from the management of particular landscapes to the management of social-ecological *processes*. From a management perspective this approach suggests that decision-makers pay greater attention to the systemic causes of change, risk, and vulnerability—factors that are quite often implicated in policies promoting increased wealth and profit opportunities for stakeholders in urban and exurban settings. Analyzing the AVI also means closely assessing the various ways the simultaneous production of risk and profits is concealed within mainstream fire and urban development discourse. This conceptual tack will entail a tight analytic interweaving of policies, social norms, economic incentives, and environmental changes that produce both increased profits and increased risks in areas currently recognized as the WUI.

Of course it would be unwise and irresponsible to just do away with the WUI altogether. The wildland-urban interface can certainly function as another useful organizing principle. Yet it does have its analytic limitations. It characterizes a land designation and set of material conditions that are grounded in a particular time and space. The WUI reflects socio-ecological changes; it doesn't produce them. (There are cases of course where a WUI designation will itself do work in the policy realm, such as with staggered insurance rates, but from a materialist perspective the WUI is an artifact of land use policies and complex planning histories.) And while it is possible for social-ecological transformations within the WUI to be tracked and measured sequentially to convey change over time, *the inadequacy of the WUI as a concept lies in its inability by itself to reveal the forces behind its own creation*. The AVI, on the other hand, is valuable for explaining complex economic, social, and environmental drivers—across multiple spatial and temporal scales—that inform the development of the WUI. Thus if we want to *change* the WUI (and not just describe it or create policies that only address symptoms and not root causes) then we need to supplement WUI analysis with dedicated, historical, and political economic AVI analysis. This marks a shift from the management of wildland-urban interface *areas* to the management of affluence-vulnerability interface *processes*.

SCHOLARLY FOUNDATIONS AND ANTECEDENTS: URBAN DEVELOPMENT, VULNERABILITY, AND HAZARD

This critical assessment of the Tunnel Fire and the concomitant production of social vulnerability and immense wealth in Oakland and other WUI areas of the American West is influenced analytically by a number of political ecologists, environmental historians, and others in related fields. There are simply too many formative and groundbreaking studies and scholars to list here (though many are referenced in later sections). There exist, however, a few productive research areas and trends that are central to this book's framing of development, vulnerability, hazard, and nature. These indispensable contributions are briefly outlined in this section. One notable area of scholarship comprises the incisive work of researchers working diligently to illuminate the relationship between environmental change, economic development, and social vulnerabilities in urban and urbanizing areas. Many of these insights emerged during the mid-1980s and early 1990s as researchers such Kenneth Hewitt, Diana Liverman, and Ben Wisner critically identified how interactions between political economies of resource use and normative planning activities came to influence which places and populations become vulnerable.[12] These and other formative researchers spurred a whole new generation of scholars and writers seeking to reveal the political and socially uneven ways in which vulnerability is produced over time.[13] Collectively these studies have shown that effective vulnerability mitigation first requires studying what Danish Mustafa labels dynamic "hazardscapes"[14] through a form of analysis described by Neil Adger as a robust assessment of "the cumulative progression of vulnerability, from root causes through to local geography and social differentiation."[15]

Being able to define, describe, and deconstruct such hazardscapes in urban and urbanizing environments requires that we first maintain a willingness to dissolve the conventional binaries of human/nature and city/rural in favor of dedicated relational geographic and historical analysis. This requires embracing (1) the mutually constitutive relationship between cities and diverse economies, policies, and ecological systems across temporal and spatial scales;[16] (2) the expansive ontological boundaries of urban nature, which as Bruce Braun notes exist as "material *and* narrated, ecological *and* political"[17]—implying that how we come to know urban environments (in ways that are often politically motivated) is intimately, and iteratively, connected to how we manage and experience them (in ways that are often manifest inequitably); and (3) the idea that cities are complex ecological systems governed to a large extent by an unremitting commitment to economic growth.[18] In this way, and following David Harvey, a dialectical perspective (inclusive of the relationship between capital, space, and nature) must be leveraged in order to reveal social and environmental changes resulting from discursive and material social relations embedded in a capitalist system.[19]

The state of California and the American West more generally during the mid-nineteenth to mid-twentieth centuries—characterized by rapid economic expansion and massive population growth—represent precisely this type of capitalist development environment. Accordingly many have used the aforementioned political ecological framework (though not explicitly naming it as such) to assess and illuminate trends in urban growth and their social and ecological implications around the Golden State and the West. These include texts such as Marc Reisner's *Cadillac Desert: The American West and Its Disappearing Water;* Gray Brechin's *Imperial San Francisco: Urban Power, Earthly Ruin;* and Stephanie Pincetl's *Transforming California: A Political History of Land Use and Development.* Each of these groundbreaking texts describes institutions and policies that have increased access to environmental resources in order to facilitate economic development around the region—nearly always with deleterious social and environmental consequences. One need only look at influential texts such as Raymond Dansman's *The Destruction of California* and Samuel Wood's *California Going, Going* from the 1960s to see early concerns over the rise of capitalist development imperatives and the production of ecological risks and social vulnerabilities across sectors, populations, and geographies of the American West. In the pages and chapters that follow I extend these important insights to the persistent, controversial, and sobering matter of development and wildfire in urban and urbanizing areas.

Vulnerability itself is a complex concept that has received considerable attention from researchers in public and private sector settings, whether in the context of wildfires or other precarious or threatening conditions (be they political, economic, or ecological). Research clarifying the meaning of vulnerability has been particularly helpful given the prevalence of related concepts such as "risk," "sensitivity," "adaptive capacity," and "exposure." Vulnerability is customarily understood as being constituted by a set of conditions or experiences that are embodied by the subject in question. For example, individuals without the social and financial resources to absorb and respond to a wildfire (or an economic crisis, etc.) hold a high level of vulnerability to that perturbation, even if exposure to the hazard in question is more or less the same for multiple people. If two neighbors have equal *exposure* to an impending wildfire sweeping over the ridge above them—both occupy identical tract homes in the path of the fire— each family will carry an equal risk of its home burning down. However, the level of vulnerability will differ depending on each household's *sensitivity*, which may include disabilities or health impairments that influence how an external stress is experienced. Similarly, household vulnerability reflects the level of insurance, rainy-day fund size, and available resource base (from family and friends) that can be readily accessed in the days and months after the fire. These may be understood as *adaptive capacity* characteristics that may ease or challenge a household's recovery process.

Accordingly, Blakie, Wisner and colleagues summarize this perspective and note that different households hold a unique ability to "anticipate, cope with, resist, and recover from the impact of a natural hazard."[20] Early political ecology analysis by Michael Watts similarly outlines the unevenness of vulnerability as it is produced and manifest within and across different scales.[21] Ray-Bennett follows suit and acknowledges that vulnerability is unique, embodied, and differentiated between individuals and communities, thus illustrating "why hazards affect people in varying ways and why people experience disasters differently."[22] Throughout this book *vulnerability* is defined as the potential for reduced well-being and lost property or life, coupled with low and incommensurate levels of potential financial reparation or other direct compensation. *Risk* is defined as the likelihood of a region, location, home, or individual being exposed to the adverse effects of fire and its associated negative externalities.

The pages that follow in this book offer much more than a narrative focused on a single wildfire event. Rather the Tunnel Fire serves as an analytic starting point for a much larger (and longer) story demonstrating the relationship between the production of vulnerability, wealth, and this dramatic event. Crucial to this longitudinal study then is the understanding that vulnerabilities shift, deepen, and in certain cases dissipate over time. This perspective has been most evident in disaster environmental histories such as Jared Orsi's *Hazardous Metropolis: Flooding and Urban Ecology in Los Angeles*, as well as in studies illustrating how access to decision making in planning influences the formation of vulnerable communities, a process effectively illustrated in Matt Gandy's *Concrete and Clay: Reworking Nature in New York City*. Perhaps most notable and relevant here is the work of Mike Davis and his illustration of the cross-temporal and multiscale production of social vulnerability in the Los Angeles area. Here Davis develops an evocative narrative depicting LA as a site for rampant development, where again and again "market-driven urbanization has transgressed environmental common sense."[23] By integrating analyses of the historical drivers of both social and environmental change, researchers such as Davis have shown that interactions between political economies of resource use and normative planning and management interventions—across local to global scales—influence levels of social stratification and ultimately which places and populations are made vulnerable.

SLOWING DOWN THE SCIENCE, CRITICALLY ASSESSING CHANGE, AND CHALLENGING CONVENTIONAL WISDOMS

Critically assessing the affluence-vulnerability interface presents an opportunity to shift from what the Belgian philosopher Isabelle Stengers describes as "rapid science" to "slow science."[24] Slow science marks a deliberate and reflexive form of scientific inquiry and is a response to the increasingly rapid nature of much research premised on producing as many findings in as short a period as possible for the purpose of responding quickly to new developments and meeting the extensive and expeditious data requests that accompany private industry–driven knowledge demands. To be sure, there is a need for fast science, particularly in the context of previously unforeseen health crises and rapid environmental changes requiring new inquires and the creation of quick policy recommendations.

But rapid science also has its drawbacks. It is largely a by-product of modern, neoliberal university and research structures requiring accelerated publication and dissemination rates, that is, increased production and efficiency outputs—from a quantitative standpoint—that are driven by institutional budgets increasingly tied to research funding procurement and publication rates. As a challenge to fast science (and neoliberal imperatives of production efficiency and knowledge immediacy) slow science reflects on queries such as: *Why are we asking certain questions? What are the implications of these questions for various interest groups? Are there concealed or opaque explanations to the problems we seek to address? What questions allow us to unearth these alternative problem framings?*

On the one hand, studying the AVI entails leveraging snapshot assessments of places and populations (i.e., examining the WUI), which generates questions such as *who* is vulnerable, *what* does their vulnerability look like, *where* are the most vulnerable groups located, and *when* do groups experience elevated risk. These who, what, where, and when questions are indicative of (though by no means limited to) fast science research that may seek to know, describe, and map the effects of fire hazards in order to generate lessons learned and to reduce risks before the next round of fires (or other perturbations) hit other, similar communities.

On the other hand, studying the AVI entails—indeed necessitates—embracing slow science. Slow science approaches to knowledge formation are less influenced by rapid publication cycles, the rush to define direct cause and effect, or structures of knowledge production shaped by private or state interests.[25] In contrast slow science concerns itself with *why* certain populations are vulnerable and *how* such conditions developed over time. These how and why questions do not prevent researchers from describing static conditions (who, what, where, when); they just present researchers (as exemplified here) with an opportunity to augment fast science research objectives with critical examinations of dynamic social and ecological processes. Unearthing complex, variegated, and systemic causes of vulnerability, and their relationship to affluence, can challenge normative conceptions of how risk and vulnerability are produced in diverse settings—not only by understanding places better but also by understanding complicated histories, knowledge networks, and spatial relations better.

Indeed the past is prologue, and in order to develop solutions to contemporary fire-related problems we must first excavate their core foundations and historical root causes.

FLAME AND FORTUNE: THE PATH AHEAD

Part I introduces the reader to important themes, places, and events appearing throughout *Flame and Fortune*. Chapter 1 has described the importance of an affluence-vulnerability interface (AVI) analytic approach as a complement to more conventional WUI analysis, while also presenting important background information on the book's primary case study thread—the Tunnel Fire. Chapter 2 establishes a conceptual justification for the implementation of an AVI approach to manage current and prospective suburban landscapes—indeed a major characteristic of the West is the immense amount of land currently still eligible for suburban and exurban conversion. Along with this important land characteristic, the chapter provides a synoptic view of the rapidly transforming West more generally through a discussion of recent suburbanization, climatic, and fire activity trends. Most importantly "the Incendiary" is introduced as a metaphor for treating the suburban West like a troubled patient (an arsonist) with deeply held and engrained behaviors and characteristics. This chapter suggests that engaging the West as merely a flammable landscape is to confront symptoms of the Incendiary, while confronting the Incendiary itself is to treat the essential character and core mechanisms driving growth and social-environmental changes in high fire risk landscapes at the urban fringe.

Part II focuses on the historical development of the Oakland Hills area (and eventual site of the Tunnel Fire). Chapters here use the Tunnel Fire area to illustrate the complex nature of the Incendiary while also demonstrating the benefits of an AVI-based analytic approach. Chapter 3 illuminates how the production of vulnerability proceeds through—and is supported by—interconnected economic development and resource use activities across city and regional scales. The chapter explores the connection between lucrative resource extraction, realty speculation, reforestation, and home construction activities in the Tunnel Fire area. These activities and resulting forms of vulnerability are inextricably linked to the development of the San Francisco Bay Area. As Gray Brechin notes in *Imperial San Francisco,* when San Francisco grows, much like all great cities, "so does both its reach and its power to transform the nonhuman world on which its people depend." He continues, "'There exists a critical *ecological* relationship between the city and the countryside, a relationship as applicable to modern San Francisco as to ancient Rome."[26] The historically resource-rich Oakland Hills "countryside" played a crucial role in shaping and indeed facilitating San Francisco's post–Gold Rush economic ascendance. These resource-provisioning activities generated roadways that several decades later fell under the speculative eye of housing developers in search of suburban homes and vacation retreats for the region's new elite. This transition from resource extraction to real estate speculation was instantiated in the landscape, as several logging paths in Oakland became arterial roads populated by municipal infrastructure, flammable tree cover, and eventually a vast collection of new home developments in high fire risk areas.

Chapter 4 continues our investigation of the affluence-vulnerability interface in the Tunnel Fire area. Specifically this chapter focuses on government retrenchment, conservative homeowner politics, and state tax restructuring spanning the 1950s to 1980s. This part of the book highlights the scalar dimensions of vulnerability-in-production. As Robert Self postulates in his groundbreaking essay "American Babylon: Race and Struggle for Postwar Oakland," the story of Proposition 13 is as much about life and livelihoods within core cities as it is about the suburbs. According to Self, Proposition 13 marked a seminal moment within California's broad ideological shift toward the pursuit and maintenance of white, neopopulist homeowner policies predicated on individual rights, estate-based wealth protection, and a nearsighted commitment to social responsibility. In the face of a postwar suburban growth politics—culminating in the overthrow, under Proposition 13, of conventional structures of taxation—metropolitan core areas like Oakland experienced tax revenue growth rate reductions, as well as depleted operating budgets within tax-dependent city fire services leading to reduced fire department budgets up to and during the Tunnel Fire.[27] In order to generate new sources of tax revenue, city officials pursued large housing developments within high fire risk areas. The gradual increase in exposure to wildfires in the Tunnel Fire area is thus deeply intertwined within California's broader tax-revolt political movement. Chapter 4 therefore challenges spatially and temporally truncated explanations of fire vulnerability that fail to grapple with complex socioeconomic factors undergirding the placement of homes in areas that are already susceptible to wildfire.

Chapter 4 ends by illustrating how factors generating vulnerability and affluence in the Tunnel Fire area also contribute to the production of vulnerabilities throughout the rest of Oakland. Minority and lower-class flatland residents experience elevated vulnerability to fires as a result of decreased fire prevention and response services stemming from the reallocation of investments and wealth into post–World War II suburban landscapes. Adding to these burdens are other potential acute impacts of city revenue curtailment where responsibility for balancing city budgets is shifted onto income and expenditure activities in poorer households—a group of city residents receiving only attenuated benefits from the region's long history of lucrative real estate developments. Here the allocation of estate-based wealth for property holders and levels of net vulnerability are highly uneven across space and demographic groups, yet also deeply intertwined in their production. This narrative thread illustrates how lucrative real estate developments and postwar tax-revolt movements are inseparable from the lives and livelihoods of inner-city, and predominantly minority, residents.

Chapter 5 provides a more nuanced depiction of vulnerability in the Tunnel Fire area. The Oakland Hills like many suburban and fire-prone areas of the West comprises residents that may not appear at first glance to be very vulnerable. The oftentimes affluent nature of these communities raises questions about what it actually means to be vulnerable given the presence of vulnerability-offsetting resources (such as insurance), the fact that risks are assumed by homeowners when buying their homes, and the possibility for homeowners to see significant property value increases over time. In light of these circumstances it is not surprising that some hold a less than sympathetic view toward residents in fire-susceptible areas, a perspective that contributed to Mike Davis's (1998) famous position on "the case for letting Malibu burn." Rather than oversimplify a diverse residential base and essentialize its vulnerability as simply a direct function of geographic location, this chapter argues for the presence of *variegated* vulnerabilities comprised in a landscape of residents, each with unique sensitivities, resources, finances, psychologies, and family histories. Findings from interviews with residents and fire survivors shed light on diverse expressions of risk and loss that vary from one individual and household to the next. The chapter argues that efforts to trivialize or ignore these risks amount to bad political ecological analysis. As part of this careful research, however, the chapter also highlights precise ways affluent communities collectively leverage their financial privileges to minimize or even offset certain risks. Well-connected individuals with the energy, means, and ability to access crucial resources are shown to have the adaptive capacity to reduce levels of risk or deliver benefits to the rest of their fellow community members.

Part III provides five examples of the complex scientific and managerial ways we come to know, govern, and debate fires in the West. Each example illustrates competing visions over what wildfires mean, how they are produced, and how they should be managed. In Chapter 6, three cases illustrate how the underlying drivers of WUI wildfires frequently mischaracterize the relative role of ecological and social structures of influence. A first case explores the rather unscientific origins of the term *firestorm* and the credibility it is afforded as a legitimate fire classification through its normative use and acceptance in mainstream fire discourse. This process diminishes the very social and profitable origins of the WUI fire problem and naturalizes these areas as a hazardous by-product of larger, exogenous, and inviolable environmental forces such as climate change. A second case examines recent efforts to study and explain the relationship between mountain pine beetles and fire activity in the western United States. As more science is generated and persistent drought conditions dominate narratives of environmental change, this evolving debate has come to rest in the familiar discursive arena of climate change adaptation. Somewhat predictably, efforts to explain problematic fire activity continue to shift from one natural, human-exempted "underlying driver" (pine beetle) to another (increased drought). A third case describes the deeply political and protracted process of challenging the economically powerful wood shingle and cedar shake industry. Wood shingles turned into political objects that seemingly came to represent a choice between the destruction of cedar shingle homes or the destruction of the cedar shingle industry. This important yet somewhat distracting public dispute has led discussions over residential fire risk to often begin not by asking *whether* to build more homes but rather by debating *how* to build them. By placing the focus of the debate on home materials and not the homes themselves, the inevitability of home construction goes largely unquestioned. Collectively all three cases illustrate how contemporary discourses on fire tend to truncate the scope of what counts (or is allowed to be brought to the debate table) as an underlying driver of increased fire activity in the West—the first two cases illustrate the privileging of climatic forces while the third describes how the development forces driving urban sprawl are treated as unavoidable and thus inevitable. As a result of these science and policy framings, management outcomes fail to address the very active, voracious, and lucrative nature of the suburban growth industry.

Chapter 7 presents two debates that illustrate how key decision-makers become mired in ideologically contentious disagreements, and how these issues distract nearly all parties from directly addressing the systemic causes of fire risk at the WUI. A first example explores contemporary debates over eucalyptus management in the Oakland and Berkeley hills. Disagreements over the flammability of eucalyptus and their nonnative status divert attention away from broader social processes: mechanisms of development that actually drive fire vulnerability (and the premise of these very debates) in the first place. The battle over eucalypts illustrates how different groups assert control over defining what belongs and doesn't belong on the landscape. In so doing, these disagreements tend to naturalize residential fire (and our concerns over fire *risk*) in the area as simply "the way things are." A second case explores yet another ideological battleground, this time pitting private property rights advocates concerned with controlling their own fire protection against those advocating for greater public agency involvement. City fire mitigation fees have produced a contentious proxy debate that forestalls other important discussions such as whether to build more homes at all and whether to shift fire mitigation efforts from adaptation to growth minimization. The placement of homes in fire-prone landscapes appears to be an unpreventable landscape reality and predetermined outcome. One of the only questions remaining (and seemingly left to fight over) is who will pay for and implement necessary mitigation activities.

These debates demonstrate the complex and controversial ways we come to know and manage fires in the West. The perspectives and disagreements found within each case study typify mainstream and scientific debates that contribute to the collective *depoliticization of fire risk and its production* in the American West. Thus just as Chapter 2 explains fire and fire mitigation as a dynamic set of sociophysical *conditions*, the five accounts described in Chapters 6 and 7 explore fire and fire mitigation as a set of contested and continually evolving *ideas*. As these cases illustrate, depoliticization sometimes occurs and is maintained by a process of repoliticization in which banal narratives, contentious debates, and the pursuit of science become mired in various alter-debates (frequently in the form of contested, place-specific issues) or proxy debates (often manifest in larger ideological disagreements).

Part IV examines the politics and outcomes of postdisaster reconstruction efforts in the fire area. Chapters 8 and 9 illustrate how many factors producing both affluence and vulnerabilities in fire areas are intensified and deepened *after* wildfires. Disaster events like the Tunnel Fire thus serve as effective catalysts for the continued growth and progression of the Incendiary. In Chapter 8, radio communication transcripts from the Tunnel Fire are utilized to illuminate specific challenges experienced by residents and responders alike at the time of the event. Based on these firsthand accounts, several important issues emerge concerning water, road, and power infrastructure. A review of reconstruction efforts in each area of concern demonstrates that progress toward reconciliation has been mixed. Capital improvements were driven largely by private property considerations and residents seeking to leverage the disaster in pursuit of neighborhood enhancements and estate-based wealth accumulation. Upgrades to water and power line equipment were lobbied and partially paid for by determined residents who used their positions of privilege to engage in collectivized risk reduction—a process described in Chapter 5 in greater detail. In these instances the community was willing *and able* to supplement beleaguered city budget capacities and help pay for municipal upgrades. This presented a win–win for residents and the city of Oakland alike. However, when private benefits were less evident (or simply not attainable)—as was the case with road-widening initiatives—residents were less apt to back such recovery efforts. As a result the pursuit of win–win outcomes unraveled.

Chapter 9 continues our discussion of postdisaster reconstruction and fire mitigation efforts. Two examples are presented that illustrate the continued extraction of profits from these high-risk areas. First, home reconstruction data in California and Colorado reveal that rebuilt homes are both bigger and more proximate to one another than prefire structures. Not only do these larger homes increase property values; they also increase overall fuel load and potential fire spread between structures (particularly in massive, high-intensity wildfires). This chapter reviews important political economic considerations leading to this reconstruction outcome that has in turn injected more fuel and value onto the landscape. Second, the emergence of the private firefighting industry is reviewed to illustrate yet another group seeking to extract profits from high-risk residential landscapes across the West. The scope and logic of this emerging sector are discussed as are its drawbacks, which include concerns by fire officials that private companies (1) lack adequate knowledge of local social and environmental conditions, (2) provide an unclear authority status that may be confusing to others, and (3) are charged with narrowly defined objectives that render mission ambivalence to other, perhaps more-exposed homes. Given the immense financial benefits and profit potentials within both the postfire real estate market and private sector firefighting industry, one is left to wonder whether the development of new suburban areas will ever slow. By the end of the chapter we are able to see that fire-prone landscapes like the Tunnel Fire area are notable for their ability to generate wealth both before and after hazard events. Over time financial opportunism has contributed to the formation of vulnerable communities while simultaneously incentivizing efforts to mitigate those very same risks. This marks a financially viable self-fulfilling prophecy: profits in production, profits in protection.

The Conclusion begins by using the Yarnell Hill Fire in Arizona to illustrate the real risks and human tragedies that arise when fighting fires that threaten residential communities. (This serves to complement Chapter 5's accounts from the Tunnel Fire.) The disaster reminds us that civil society, politicians, city planners, and private developers among others should no longer conform to fiscal pressures and incentives at the metropolitan fringe as if there were no consequences; as if real tragedies and losses will not follow. Collectively these groups must stop merely treating the symptom of the problem by putting out small fires within the WUI. Instead we need to address the root causes of the firestorm itself, by treating western WUI areas like a patient—the Incendiary—through a comprehensive assessment of their backgrounds, histories, underlying drivers, internal governing mechanisms, core characteristics, and connections to externally influential forces. Although no single template exists, the Conclusion outlines a few strategies that will help initiate a conversation about how to shift from current tactics that grapple with WUI *symptoms* to more innovative approaches that directly tackle AVI *processes*. The first two approaches, "refining and expanding community adaptation" and "fighting fire with fire," reflect more moderate and extreme strategies, respectively. The third approach, "reducing demand for development," provides somewhat of a middle-ground framework. This third approach is itself composed of three sections: disabling development, discouraging development, and deflecting development.

2 The Changing American West

FROM "FLAMMABLE LANDSCAPE"
TO THE "INCENDIARY"

"The Flammable West" is a phrase that gets used often in media reporting. An evocative label, it depicts a region that seems almost like clockwork to "catch fire" and "go up in smoke" each year. In an era of transformative climatic change and protracted drought, it is easy to understand this common portrayal of the American West. A 2013 Northern California public television news article titled "The Flammable West: Mega Fires in the Age of Climate Change" is a good example. The article provides a useful if startling description of fire trends in the western United States. It tells us that compared to annual averages during the 1970s, the period 2002–11 contained twice as many fires larger than 1,000 acres, seven times more fires exceeding 10,000 acres, nearly five times more fires greater than 25,000 acres, and an average fire season lasting 2 1/2 months longer.[1] Like many other reports and articles, "The Flammable West" provides the reader with a valuable primer on the state of fire in the region. These publications serve an important function as they educate the public and support larger efforts to introduce and implement fire mitigation and adaptation policies. But they also reproduce and fortify a troubling trend within the suburban fire discourse: the persistent focus on the region's tendency to burn, as if this were the natural order of things. As if flammability *is* the

39

problem and not the symptom of a larger, engrained, and more pernicious underlying set of processes.

TREATING MORE THAN SYMPTOMS: THE AMERICAN WEST AS THE INCENDIARY

The time has come to illuminate flammability. In both mainstream reporting and contentious debates the term *flammable* (or *tinderbox*, etc.) is often deployed in a manner that conceals, and thereby depoliticizes, short-sighted and sometimes reckless landscape change and development histories (see the Preface for an example). Flammability connotes the physical symptom of a landscape but not the root causes behind its making. Consider instead the term *incendiary*, which in *noun* form implies that an object (or person or place) is an agent that actively produces and incites fire. It *makes things* flammable, much like an arsonist. The term *flammable* on the other hand implies that an entity, such as a landscape, holds qualities that make it susceptible to fire. It is an adjective and a passive characteristic, connoting that an object just *happens to have* the capacity to easily go up in flames.

Fire can be understood as a symptom of a flammable landscape, while landscape flammability exists as one symptom of an incendiary. To combat the symptoms of flammability one will find it sufficient to focus on the WUI environment. This will entail engaging in active management that includes diverse and ongoing fire mitigation treatments and community adaptation efforts. To combat an incendiary, however, is to tackle something much more engrained and systemic to the entity in question. It is to confront its essential nature and characterological qualities. Within the American West, confronting the *landscape* as an incendiary thus requires wielding an AVI framework and engaging in historical analyses of culture, political economy, environmental politics, and social-environmental change.

Imagine a network of elusive, brazen, and dangerous arsonists afflicting a series of towns and cities around the West. Every few weeks these individuals randomly ignite one or two fires. Some of the fires are controlled with only minor damage while others quickly spread and endanger nearby communities, resulting in lost lives, considerable private property damage, and millions of dollars in firefighting and rebuilding costs. One way to handle the problem would be to simply confront the periodic fires through quick and efficient emergency response efforts. But for many this *reactive* approach might seem too limited. A more robust response might be to tackle flammability in the area head on. This would entail mitigating flammable land features through vegetation clearing and building code modifications; increasing enforcement by limiting access to particularly flammable areas; or even eliminating (or at least tracking) the sale of materials that make the combustible devices used by arsonists. These approaches fall under the category of fire *adaptation*.

While a combination of these strategies (reactive and adaptive) may appear advisable, it could be argued that a more effective and long-lasting approach would be to also directly confront the source of the problem itself. That is, investigate the incendiaries and undercut the arsonists. Why are they lighting the landscape on fire? What is motivating this destructive behavior? How are they getting the necessary money and resources? And what is it about their environment, background, character, and psychology that leads them to perpetrate such acts? To address these questions is to grapple with the root causes of the problem. To avoid these questions is to leave the arsonist cell unchallenged and intact. This approach accepts that while it is important to treat the source of fire—flammability—it is also important to treat the source of flammability—the Incendiary. Confronting the Incendiary means closely examining its history and foundational characteristics. As a society we would never accept simply reacting and adapting to an arsonist.

The same approach holds true with wildfires. If we understand the landscape as a patient, as the Incendiary, then the best way to substantively reduce the symptom of flammability is to engage in appropriate fire reaction and mitigation activities *while also* confronting the political economic structures, sociocultural behaviors, and environmental systems that continue to produce, support, and enrich the Incendiary. Unfortunately our understanding of the relationship between landscapes and fire in the American West is shaped by debates that often conceal this incendiary narrative and naturalize landscape flammability. Part of the effort to confront the Incendiary will require describing how the nature of these debates effectively depoliticizes the political histories and patterns of material accumulation that have given rise to wildfires and elevated levels of social vulnerability.

Consider the 2016 Fort McMurray fire, which burned hundreds of thousands of acres in Alberta, Canada. As the massive fire still burned, a chorus of articles covered the fire using titles such as "We Need to Talk about Climate Change: Tragedies Like the Fort McMurray Fires Make It More Important, Not Less." The article ties the massive blaze to the impacts of climate change and points out the clear and present dangers of our now drier, longer, and more disastrous fire seasons. The author notes that the cause of the fire is indeed a "messy mix of factors" including forest management practices, urban encroachment, and the effects of El Niño. But the article also singles out climate change as the topic (and causal factor) that we have failed to grapple with directly or effectively at the media and policy level.[2] The argument that we need in order to address the elephant in the room—climate change—is true to an extent. Climate change is extremely important and not adequately accounted for in many policy circles. But a quick read of fire reporting, including another article titled "Fort McMurray and the Fires of Climate Change" that makes a similar argument, leads one to wonder just how marginalized the issue of climate change really is within the media.[3]

But much more importantly, the leap to illuminate (and implicate) climate change has the simultaneous effect of concealing the important role urban expansion and lucrative developments have in creating this "tragedy." The McMurray fire would surely have received much less coverage if it seared only through the surrounding, uninhabited boreal forest. What gets overlooked in this climate-frenzied coverage is Fort McMurray's development history: rapid growth in population and size over the past several decades supporting large-scale oil extraction from an enormous subterranean tar sands deposit. When the focus is only on the fire's impacts, the landscape's flammability, or the influential role of climate change, the opportunistic actions of corporations and governments seeking to capitalize on this lucrative landscape (i.e., the Incendiary) fade into the explanatory background. City inhabitants are rightfully portrayed as the victims, but so too are the city officials and oil industry players that continue to fuel this regional growth. Moreover the fact that the Fort

McMurray area was developed *in pursuit of fossil fuels that drive anthropogenic climate change* is also rendered marginal to the story. Not only are patterns of regional development crucial to explaining wildfire hazards; they are also central to explaining the problem of climate change! If one drills down to identify the structural root causes of fire disasters like Fort McMurray, then patterns of rapacious urban development are almost certainly what will be found.

Before providing an in-depth investigation of one such complex and lucrative development history (using the Tunnel Fire area as a case example), and before showing how knowledge of these histories gets buried and inoculated beneath several layers of alternative debate topics, let's briefly review the symptom of flammability in the suburban American West.

A BRIEF CHARACTER PROFILE OF THE INCENDIARY: THE CHANGING AMERICAN WEST

Increased Fire Activity, Intensity, and Costs

Fire is a constant in the American West. Historical records show that periodic wildfires are a normal component of the region's ecology with fire recurrence intervals ranging from less than one decade to over two centuries depending on the landscape and fire intensity in question.[4] Fires are also extremely beneficial to the region. They promote ecological succession for certain species by for example helping pinecones open and assisting seed germination for many chaparral plants—all of which allows diverse habitats to regenerate and remain viable over time. Fires remove forest floor debris, open up forests to sunlight, and nourish soils. They also help ward off pests, diseases and, insects that may otherwise reduce ecological function.

However, as temperatures rise, droughts persist, and urban populations swell, fires in the region command more and more of our attention. As the aforementioned "Flammable West" report describes, large fires seem to be more frequent and devastating than in the recent past. Fully explaining this increase in size and frequency is extremely difficult, although many experts have persuasively noted the negative long-term effects of pervasive fire suppression policies that have led to significant fuel buildup in many vegetated areas (particularly near human assets) around the region.[5]

By most accounts we have a very serious problem in our midst. Consider a recent study examining large fires (over 988 acres) in the western United States. From 1984 to 2011, the number of these fires across all ecoregions increased at a rate of seven fires per year. Meanwhile total fire area increased at a rate of 87,722 acres per year.[6] Fires exceeding 100,000 hectares (247,105 acres) are now much more common than in previous decades,[7] and the length of fire seasons are now longer by several weeks and even months in some locations.[8] Moreover the influence of wildfires on human populations is evidenced by the growing number of structures damaged by wildfire. In total, from 2000 to 2012 the United States lost 38,701 structures to wildfires, an average of 2,977 structures per year.[9]

Similar unpropitious news can be found at the state level. In California since 1923 fifteen of the most damaging twenty fires (in number of structures destroyed) have occurred within the past twenty-five years; nine of these fires have occurred over the last ten years.[10] This means that in California's modern history about 75 percent of the largest and most destructive wildfires have occurred in the past twenty-five years and nearly 50 percent have taken place in the last decade alone. A comparable story is unfolding in Colorado where the average number, total area, and average size of fires have also rapidly increased since 1960. The average size of fires in Colorado during the 1960s was just under eighteen acres per fire. During the 2000s the average fire size was just under forty acres. For the same two periods the average annual area directly impacted by fire rose from just over 8,000 to 96,000 acres. The average number of fires per year also increased steeply from the 1960s to the 2000s—from roughly 260 to 2,500 fires annually.[11] Much like California and the rest of the American West fires in Colorado are larger and more prevalent than at any time in the past half century.

Meanwhile over the past fifty years the cost of fire mitigation activities has grown dramatically in the United States. In the 1970s the federal budget allocated to fighting wildfires averaged $420 million. This figure jumped to $1.4 billion by 2000[12] and increased again to $2.5 billion by 2012. It is estimated that the total fire mitigation budget in 2012 stood at a lofty $4.7 billion when inclusive of federal, state ($1.2 billion), and local

($1 billion) governments each year in the United States.[13] These costs have risen primarily as a result of increased fire mitigation requirements due to a buildup of fuels resulting in part from past fire suppression policies, a warming climate, persistent drought conditions in the West, and perhaps most importantly, the development of residential communities adjacent to already fire-prone public lands.

The potential costs for homeowners and insurance companies are equally staggering. In the western United States, it is estimated that nearly 900,000 residential properties are currently located in high or very high wildfire-risk categories (as designated by state and federal fire zone severity mapping projects). From an economic standpoint and on the basis of total reconstruction cost estimates, this presents a potential fire loss value of more than $237 billion. In the very high risk category alone there are over 192,000 residences with a potential loss value totaling more than $49 billion. The Denver, Colorado, metropolitan area for example encompasses more than 35,000 homes in very high risk areas holding an estimated reconstruction cost of more than $10 billion. This same designated area in the San Francisco Bay Area contains 5,800 homes and $2.7 billion in potential lost value, while in the San Antonio, Texas, metro area the numbers are 31,000 and $7 billion.[14]

Similar numbers can be found across the West and when viewed collectively seem to portend financially catastrophic outcomes. However, a more comprehensive assessment will also reveal the immense revenue potential lying within these flammable landscapes. Payouts for postdisaster reconstruction materials and labor make these areas (and their inevitable fire events) a lucrative destination for home construction industry manufacturers, local vendors, and private fire service outfits. Moreover this form of disaster capitalism does not just simply lead to the reconstruction of homes; it results in the development of bigger, better, and more valuable homes.

Suburbanization in the West and the "Room for Growth" Problem

The post–World War II suburban landscape in the western United States has undergone a complete home makeover. It is a transformation *by* homes, not *to* homes. Population growth and sprawling residential developments have increased human encroachment into fire-prone areas. Stephen Pyne, a foremost expert on fire history, has noted that even as early as the mid-1950s the West had transitioned into a "new fire regime." No longer primarily confronted with traditional forest fires, firefighting industry participants were now asked to tackle a "lethal mixture of homeowners and brush."[15] Similarly the urban historian Mike Davis—a wry and incisive critic of WUI residential developments—remarked that "the artificial ecotone of chaparral and suburb magnified the natural fire danger while posing new obstacles to firefighters, who now had to defend thousands of individual structures as well as battle the fire front itself."[16]

Recent population growth in the West is unmistakable. Between 1980 and 2006, eight of fastest-growing eleven states in the United States were located in the West. These include Nevada, Arizona, Utah, Alaska, Colorado, Idaho, Washington, and Colorado. Over the same period the thirteen states comprised in the American West witnessed a growth in population from 43.2 million in 1980 to 69.3 million in 2006. This marks an increase of over 26 million people and a growth rate of 60.5 percent, far exceeding the 25.5 percent population growth experienced in the rest of the United States. By 2015 the region's population exceeded 76 million.[17] Substantial demographic changes have also occurred, including a drop in the percentage of white residents from nearly 90 percent in 1980 to roughly 60 percent in 2006. Meanwhile Hispanic populations increased from 14.5 percent to 25 percent of the total population. The African American population increased by 31 percent in the West while Asian American populations increased by 138 percent.

Urban areas in particular saw dramatic growth during this period. Metropolitan regions of the West experienced a 63.6 percent increase in population compared to 36.8 percent for nonmetropolitan areas.[18] Both of these changes far exceed the pace of urban growth in the rest of the United States. They also reflect the West's rapid suburbanization. While recent efforts have been made to engage with urban in-fill practices (i.e., by making cities more dense through sustainable city initiatives such as Smart Growth panning and New Urbanism design principles), the western United States is unmistakably more suburban than it was during the middle 1900s. Thus urban expansion in the West is occurring in the form of both higher population totals and increased geographic size.

Suburbanization around the region has increased the number of houses in urban areas by as much as 27 percent from 1970 to 2000, with approximately 38 percent of this new development occurring near or within the WUI.[19] Between 1990 and 2000 alone, more than one million homes in total were introduced to the WUI in the states of California, Oregon, and Washington.[20] Across the western United States, WUI areas have seen a 300 percent population growth rate in the past fifty years, which outpaces overall regional population growth rates for the same time period.[21] In geographical terms these areas in the West have experienced 60 percent expansion since 1970;[22] traditional wildlands were converted to areas characterized as the wildland-urban interface at a rate of four hundred acres per day, an equivalent of close to two million acres per year.[23]

The most alarming suburbanization statistic, however, concerns what *hasn't* been developed. As of 2008 only 14 percent of private land in WUI areas of the western United States had actually undergone land conversion. By 2013 this number increased to 16 percent (see Figure 5).[24] While these statistics may at first appear to be a silver lining (i.e., patterns of sprawl and land conversion are not as bad as we might have imagined), these numbers actually reveal something quite startling: over 80 percent of the WUI environment remains eligible for further growth, greater levels of social vulnerability, and increased firefighting costs. Nationally a study by the Forest Service estimates that 21.7 million acres of rural land within ten miles of the national forests and grasslands in the lower forty-eight states will experience increased housing development by 2030. As of 2012 forty-six million homes were located in the WUI. On the basis of current trends that number is expected to increase to fifty-four million by 2022.[25]

The possibility for significantly more land development at the urban periphery should serve as a clarion call for more coordinated action by governments, private property owners, and nonprofits alike. Although many politicians express concern about rampant metroplex sprawl into surrounding areas, the fact remains that things could get much worse in the years and decades ahead. Given these conditions it seems now is the logical time to stem the tide and reduce further sub- and exurban developments around the West. Of course these development trends will only be reversed if we pause and treat the Incendiary rather than focus on its flammable symptoms. This entails engaging the AVI while avoiding the

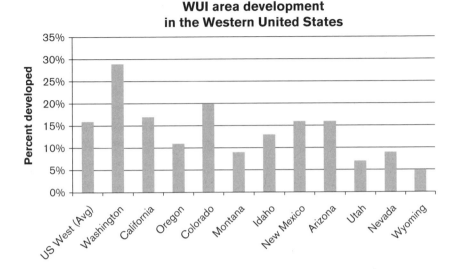

Figure 5. Percentage of WUI area developed in the western United States as of 2013. (Source: Headwaters Economics)

depoliticization and repoliticization trap that naturalizes, legitimates, and victimizes home developments across the West.

The American West's Changing Climate: Adding Fuel to the Fire

Human exposure to wildfires and the social and economic costs of fighting them continue to rise in the West. The primary explanation for this increased risk is persistent urban expansion, its encroachment into already-fire-prone environments, and the general conversion of rural, agricultural, and wildland landscapes into suburban and exurban communities. With this encroachment comes growing human exposure to wildfires. It is now understood for example that human population density is a more significant predictor of fire activity than forest density.[26] As previously mentioned, wildfires are common occurrences in the West even in the absence of human activity, due to normal climate variability and frequent and sometimes prolonged droughts. Wildfires have occurred for millennia and provide crucial ecological services required to recycle nutrients, improve soil condition, and initiate plant succession.

In the western United States topographic features influence moisture availability and temperature and therefore the composition and structure of plant communities. As a consequence fire regimes supported by these different plant communities change across the landscape. Consider some portions of the Front Range of Colorado where wildfires burn frequently (about every ten years) in low-elevation pine forests due to higher evapotranspiration rates, and much less frequently (about every 150 years) in higher-elevation subalpine forests due to greater latent moisture levels.[27] Although the West has always experienced routine and sometimes large wildfires (and although these trends are increasing due to suburban development), fire activity is also changing because of a dramatically altered western climate, a climate now characterized by higher regional average temperatures, increased rates of evapotranspiration, and more pronounced levels of aridity (at least compared to recent history). These emerging conditions are in turn resulting in longer and more active fire seasons.[28]

The Fifth Report of the Intergovernmental Panel on Climate Change (IPCC), released in 2014, sheds crucial light on the changing western climate and its implications on fire activity. According to the IPCC,[29] western North America "has experienced drier conditions since the 1970s," is experiencing "earlier peak flow of snowmelt runoff in snow-dominated streams and rivers," and is witnessing "declines in the amount of water stored in spring snowpack in snow-dominated areas."[30] Less precipitation falling as snow and an accelerated melt rate for remaining snowpack are in large part a symptom of rising average temperatures in the western United States. From 1970 to 2013 average temperatures across the region increased by 1.9 degrees Fahrenheit. This is a substantially higher increase in average annual temperature than in other regions of the United States.

Collectively these factors have produced a fire regime that is significantly longer than previous decades. Today for example the western wildfire season regularly runs seven-plus months, which is much longer than the typical five-month-plus season experienced during the 1970s.[31] While winter fires in more arid regions of the West such as Southern California are unusual, historical records show they do occur from time to time. For many other parts of the West, however, these changes are requiring new response and adaptation measures. For example the 2014 Red Fire, which raged near the city of Arcata in far Northern California, had fire managers scratching their heads—not because of the size or severity of the fire but because it seared the landscape in early January, a time of year when this coastal area normally contains high humidity and soggy ground conditions. The Red Fire prompted an unusually busy Marin County battalion chief to note, "It's unprecedented for us to do this in January."[32]

Meanwhile, according to a Northern California fire marshal, extended fire seasons have far-reaching impacts for all parties. Speaking in late 2014, she mentioned that "our guys never took their wildland gear off the engines last year . . . usually you'd take them off and they're done, they're off right in November; they didn't take them off last year." As residents and fire departments scramble to meet these changing needs, the total number of annual large fires continues to grow. According to the Union of Concerned Scientists, the number of annual fires greater than one thousand acres has increased from approximately 140 in 1980–89 to 160 in 1990–99 and again to 250 over the period 2000–12.[33]

Climate projections for the West indicate even more fire activity in the coming decades. Large fires under future climate change projections are expected to increase by up to 175 percent in the Rocky Mountain region from 2000 to 2050, for example.[34] The comprehensive 2014 IPCC report offers still more sobering news. As various physical and social aspects of climate change are described, each assessment and projection is assigned a different confidence level. (These are based on the ability of scientific reports to track and explain one phenomenon in relation to another.) According to the IPCC report, there is "high confidence" that "wildfire-induced loss of ecosystem integrity, property loss, human morbidity, and mortality" will emerge as a result of increased drying and temperature trends in the region. The IPCC also assigns a "level of risk and potential for adaptation" to various climate change impact categories such as flood and fire. According to the report, wildfires in the United States will pose a "highly significant" future risk. Long-term (2080–2100) wildfire-induced losses due to increased drying and temperature trends are given a risk rating of "medium/high" under a 2 degree Celsius increase scenario, and a "high" rating for a 4 degree Celsius increase. To put this in perspective, this is a higher risk score than nearly every other category across all global regions and threats, save for coral reef damage in Australasia and associated declines in biodiversity and fisheries abundance in ocean reef environments. Of all the global risks examined under climate change, risk to wildfires in the western United States is one of the most sensitive to increased future temperatures.

But to be clear, while climate change itself is certainly generating environmental conditions favorable to higher-frequency and -intensity fires, it is the region's long history of fire suppression and most importantly the widespread encroachment of human populations into already high fire risk areas that are most responsible for increased fire exposure, risk, and mitigation costs across the region. The effects of climate change on the American West are in many ways a lot like adding fuel to an already burning fire.

PART II Illuminating the Affluence-Vulnerability Interface in the Tunnel Fire Area

3 Trailblazing

PRODUCING LANDSCAPES, EXTRACTING
PROFITS, INSERTING RISK

By three o'clock on the afternoon of October 20, the Tunnel Fire had seared through roughly two miles of hilly residential communities to the south and southwest of its epicenter. Carried by hot, dry Diablo winds (the Bay Area's equivalent of Southern California's notorious Santa Ana winds), the fire proved both tenacious and voracious. The flames were, as the Battalion 2 chief put it from his command post, "completely out of control on several fronts." Though moving primarily in a southerly direction by afternoon, steep topography adjacent to the firestorm's origins and periodic winds from the east propelled the fire nearly two miles due west toward the San Francisco Bay. In the fire's path loomed the iconic Claremont Hotel and Resort.

The Claremont Hotel, which stands today as a regional landmark and visually dominating icon at the foot of the Oakland and Berkeley hills, was constructed alongside resource extraction access roads in 1915.[1] Built by a small group of local real estate developers, the hotel was erected primarily to attract home construction investments and potential homebuyers to hillside areas just upslope from the sprawling hotel complex (see Figures 6 and 7). The hotel has thus come to represent the conjoining of two periods of regional economic development: first, resource extraction activities

Figures 6 and 7. The iconic Claremont Hotel (top image, right foreground) sits below the northwesterly edge of the destructive and fast-spreading firestorm. The hotel approximately seventy-five years earlier (bottom image) sits amid drastically different land cover. (Credit: UC Berkeley Bancroft Library)

through its placement at the base of former logging roads, and second, local real estate development activities by serving as a logistical hub for the establishment of home construction syndicates connecting these old logging routes to the region's once popular Key Route mass transit rail system. In this way the Claremont Hotel signifies a transition in the Oakland and Berkeley hills from a productive resource extraction area to a valued landscape falling under the speculative eye of housing real estate developers.

On the afternoon of the fire, gusting winds and swirling smoke obscured views of the hillside directly behind the hotel. Occasionally lapses in wind would produce vertical convection columns revealing large orange flames whipping high into the air just above the hotel (see photo on cover). With 279 guest rooms and an attached twenty thousand square-foot spa, the hotel signified an immense fuel load that could potentially propel the fire further into nearby communities and toward the University of California Berkeley campus. With this concern in mind the hotel was targeted by firefighters as a high-priority fire perimeter. A number of fire units were stationed at or nearby the hotel grounds. After a long and arduous fight that demanded a significant amount of mutual aid, the Claremont Hotel was saved from the fire—a salvation many firefighters admit was more a by-product of flagging winds than active fire mitigation efforts. Meanwhile the very homes the Claremont Hotel had helped market to prospective residents over half a century earlier burned to the ground just upslope. Vegetation that was used to attract those homeowners to the area also suffered extensive damage and in some locations even assisted the fire's rapid spread.

Three quarters of a century after its construction, the Claremont Hotel and Resort edifice stood intact, tucked safely beyond the fire's most northwesterly reach. Meanwhile the landscape it once helped transform, market, and populate was left in ruins. In the aftermath of the firestorm the Claremont Hotel and its environs has come to represent more than just a landmark building saved from destruction. Rather, in a single structure it effectively illuminates three historical and geographical processes—extensive land cover change, large-scale real estate developments, and extensive regional suburbanization—that have led to the concomitant production of wealth and risk, not only in this Northern California hillside neighborhood but also around the San Francisco Bay Area, the western United

States, and beyond. In order to fully understand the Incendiary and the root causes of suburban fire risk (and to counter the persistent depoliticization of these explanations), it is necessary to fully engage with this complex and elucidatory history.

TIMBER EXTRACTION, ROAD INFRASTRUCTURE, AND DEVELOPMENT "MOMENTUM"

This description of vulnerability, affluence, and their coproduction begins with large-scale clear-cutting of coast redwood (*Sequoia sempervirens*) occurring in the area as early as 1840. This sharp upswing in timber extraction supported an increasingly ravenous Bay Area timber industry as the city of San Francisco's population grew from roughly 30,000 in 1850 to nearly 350,000 by the turn of the century. During the same period the city of Oakland expanded from a newly incorporated town of only a few hundred to roughly 70,000 residents. Oakland's growth was aided in part by its selection in the 1860s as the terminus of the transcontinental railroad. Meanwhile as San Francisco and Oakland grew, so too did the appetite for water, lumber, and other natural and mineral resources from the now rapidly expanding region.

Diverse and in some areas dense tree cover in the Oakland Hills was supported by the undulating topography of this inner coastal hill range, including its many deep canyons and ravines. While much of the original redwood forest occurred a mile or two south of the Tunnel Fire area, the influence of logging activities on subsequent community developments in high-risk fire areas is unmistakable: these early logging activities eventually led to the placement of municipal infrastructure that would, several decades later, be used by landholders, developers, and city officials to justify the placement of new homes in areas historically exposed to wildfire.[2]

During and immediately following the Gold Rush boom cycle of the mid-1800s "redwoods as tall as 300 feet and as wide as 32 feet" were captured and hauled to shipping points and sent for home and commercial construction purposes in San Francisco and Oakland.[3] This lumber was also used to build Mission San José and replace structures and resurrect portions of the burgeoning city of San Francisco that had burned during major fires in 1850 and 1851. Moreover the vast body of human labor behind these extractive activities occupied makeshift structures in shantytowns, which were also constructed out of timber from the East Bay hills. In this way both the material ends and means of regional economic development during the mid-1800s were made possible by this vast supply of East Bay timber.

Redwood trees in particular are an extremely valuable source of lumber. They are saturated with astringent tannin, meaning that fungi, pests, and fire rarely penetrate their thick bark. The redwood's resistance to deteriorating environmental conditions increases its growth longevity leading to strong, durable, and straight lumber. As one naturalist put it, citing a free-market reductionist view of this massive tree species, "One old-growth redwood tree can make enough lumber to build five eight-bedroom houses."[4] While this might be a slight exaggeration, the rapid increase in value of redwoods is hard to overstate. Consistent with basic supply and demand logic, commercial redwood lumber prices reflect increased demand (and quickly decreasing supply) as prices skyrocketed from $30 per one thousand board feet in 1847 to upward of $600 in 1849.[5] By 1852 there were four steam sawmills operating in the Oakland Hills (see Figure 8). The operation was so extensive that by 1860 hardly a tree remained.[6] In lower hill areas various farming activities including wheat and hay production ensued during subsequent decades. Many of these grains were used for feed within newly established cattle-grazing outfits in the foothills region—an emerging set of land use practices that utilized the recently denuded landscape now free of coast live oak (*Quercus agrifolia*), redwood, pine, and other native riparian woodlands.

The role of the East Bay hills as a critical source of raw materials for the construction of San Francisco during and immediately following its Gold Rush economic ascendance should not be underestimated. The Sierra Nevada foothills contained mineral deposits paving the way for wealth generation and frenzied infrastructure investments in the San Francisco Bay Area during the middle 1800s. Meanwhile the East Bay hills enclosed a ready supply of raw construction materials to facilitate the actuation of those investment objectives, particularly during the 1840s (and less so in the 1850s); it was a thrust of development that in turn spurred a regime of continued material accumulation for several decades to come. For its

Figure 8. An 1880s sawmill in the Oakland Hills along Palo Seco Creek. After the removal of nearby redwoods, other tree species were also felled and hauled downslope. (Credit: UC Berkeley Bancroft Library)

ability to serve as a major source of timber, mineral (e.g., pyrite, sulfur), and grazing resources, the East Bay hills region stands as an early (and often overlooked) site of instrumentalist resource extraction supporting frenzied construction activities in the San Francisco Bay Area during its early rise—a role later played by the likes of California's Central Valley, Hetch Hetchy Valley, and North Coast forests.

A trajectory toward increased fire risk in the Oakland Hills was set in motion as a result of this cross-bay relationship. Logging, resource extraction, and land conversion activities during the mid-1800s not only altered the Oakland Hills landscape; they also introduced crude infrastructure and laid the foundations (and momentum) for subsequent housing development trajectories (see Map 4). Many current-day roads that abut or cross through the Oakland Hills fire area[7] originally terminated at logging sites and were used to haul timber downslope to the Oakland Estuary and across the bay to San Francisco. Several other East Bay roads[8] also began as logging roads.[9]

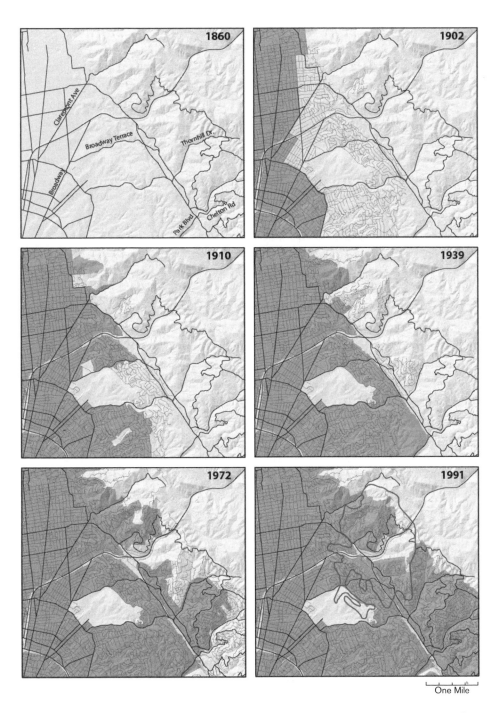

Map 4. Map sequence illustrating development of road and home construction in the city of Oakland. Real estate developers utilized early logging roads in subsequent decades as preestablished and cost-efficient points of entry into the hillside. Shaded area indicates extent of home development, which generally follows road construction. (Credit: Eric Ross)

Even arterials to the north of what is generally understood as good lumber areas were utilized as part of the East Bay logging enterprise—roadways that today cross directly through the 1991 Tunnel Fire zone. According to one historical account, "The Oakland lumber yards needed in 1852 a direct route to the woods. . . . so a road was built that followed, very roughly, Broadway, Broadway Terrace and Mountain Boulevard."[10] While many of these roads have been significantly widened and modified, they remain in the same graded locations as they existed during the mid-to-late 1800s.

AFFORESTATION, EUCALYPTUS, AND THE PRODUCTION OF DESIRABLE LANDSCAPES

The Claremont Hotel signifies a transition in the Oakland Hills from productive logging area to a valued landscape falling under the speculative eye of housing real estate developers. The introduction of crude logging road infrastructure generated eventual access possibilities for developers, who in subsequent decades constructed thousands of homes in areas that would witness scores of nearby and adjacent wildfires. Ensuing tree importation and afforestation efforts would further facilitate housing tract speculation, construction, and marketing efforts.

Early tree planting can be traced to the well-known essayist Joaquin Miller, who in 1886 purchased sixty-nine acres and immediately planted pine, cypress, acacia, and eucalyptus on his property.[11] Figures 9 and 10 depict the denuded landscape confronting Joaquin Miller and his wife, Abigail Leland, immediately after the construction of their homestead (known to downslope residents as the "Heights") and the subsequent rapid afforestation around the Miller homestead. While that homestead was more ornamental in nature and essential to water retention and the stabilization of adjacent hillslopes, others were more entrepreneurial in nature and engaged opportunistically in forms of early agroforestry. Before major housing developments took hold, these early tree farms blanketed much of the area. This marked a crucial, indeed dramatic, ecological transformation in the area: within a single generation humans removed nearly all local vegetation cover and introduced a completely new mosaic of tree and shrub species.

Figure 9 (top). Intensive logging during the mid-1800s left much of the Oakland Hills area void of tree cover. The Joaquin Miller homestead (1886) sits in the foreground. (Credit: UC Berkeley Bancroft Library)

Figure 10 (bottom). By 1913 the environment around the Miller homestead and much of the Oakland Hills had changed considerably. Nonnative and ornamental species can be seen around the property. (Credit: UC Berkeley Bancroft Library)

This ecological transition began with an influx of seeds and public and private money. From 1885 to 1893 these early forestry efforts were guided by the California State Forestry Board, which favored quick-growing eucalyptus trees. Between 1910 and 1913, nearly fifty years after the removal of virtually all tree cover in the hills region, Frank Havens, a prominent landowner in the region and an associate of the East Bay Watershed Company, planted approximately three million nonnative eucalyptus (*Eucalyptus globulous*) and Monterey pine (*Pinus radiata*) seedlings along the region's hillslopes. Eucalyptus trees were planted for commercial lumber speculation due to their purported lumber quality. During the next several decades, the fast-growing blue gum eucalyptus indeed proved to be a hearty species tolerant of high winds, shallow soils, and seasonal drought conditions—an ecological temperament perfectly suited for the East Bay hills.[12]

Unfortunately for the timber industry eucalyptus also turned out to be a poor source of construction-worthy lumber due to its braided wood fibers, uneven grain, and tendency to split and warp after drying. Although they had a penchant for growing quickly, eucalypts were incompatible with another intended purpose: the construction of railroad ties to serve the Union Pacific Railroad and the anticipated arrival of other regional and transbay rail systems similar to the Key Route system. The wood tended to twist while drying and was so hard that hammering rail spikes proved extremely difficult. It also became increasingly undesirable as an energy source as cheap oil overtook demand for wood fuel.

Despite having low construction utility, the eucalyptus still served a crucial development function. Beginning in lower hills areas, these tree species were valued for their ability to beautify and increase land values in the hillside by replacing unsightly barren slopes with a bucolic residential environment suited for affluent members of the San Francisco and East Bay business class.[13] As Figures 11 and 12 illustrate, real estate developers lined roadways and filled properties with eucalypts as an explicit marketing strategy. While eucalypts may have been a poor source of construction-quality timber, they remained a vital component of residential construction in the region more broadly—repurposed and revalued in their upright form as an elegant, shade-bearing, and sweet odor–producing tree cover.

Figure 11 (top). Post-logging replacement cover arrived in the form of eucalyptus trees and other nonnative and ornamental species. Here, a eucalyptus farm is placed alongside a former logging road. (Credit: Oakland Public Library History Room)

Figure 12 (bottom). Many eucalyptus saplings were sold and replanted in various locations. Others, as shown in this real estate brochure, remained in place and contributed to the roadside aesthetics of new real estate tracts. (Credit: Oakland Public Library History Room)

Momentum toward increased home development—influenced in part by the introduction of hillside logging roads—proceeded in two ways. On the one hand, graded and reinforced thoroughfares were viewed simply as a cost-efficient means of bringing roads to large real estate owners waiting to initiate home construction. The City of Oakland and local banks providing loans to public service corporations held a favorable view of road and municipal infrastructure financing that reduced outlay costs. Permitting new residential developments was simply easier and cheaper in areas that lay adjacent to preexisting haul roads. On the other hand, many hillslope property owners—especially those residing nearby but not necessarily adjacent to major logging roads—were not content to sit idly by waiting for infrastructure to come to them. Powerful landowners such as the Realty Syndicate indicated they were "not compelled to wait, as is the individual, upon the completion of corporation or municipal facilities" as they proceeded to "develop neighborhoods of a high class nature in absolutely new districts."[14] Under this more proactive approach, large real estate holders improved their property through leveling, tree planting, and basic roadway construction activities prior to the completion of more robust city-funded residential road improvements. These initial investments, land improvements, and makeshift spur roads helped to attract further public investments and "pull" housing developments (and properly graded roads) onto their property.

Following these successful eucalyptus and pine plantings, directors of the Realty Syndicate, including Frank Havens and his associate F. M. Smith, noted "the increased value their holdings would have if plentifully timbered."[15] The Mahogany Eucalyptus and Land Company, which dominated property holdings at the turn of the nineteenth century in what is now the Tunnel Fire area, noted that "this tree at this particular moment is in many instances the most valuable one on the face of the globe.... The Company now sees plainly that it possesses a source of emolument higher than that of the average gold mine."[16] Half a century earlier one would have assumed the tree species receiving such praise was the mighty redwood, replete with its abundant supply of dependable and high-grade timber. But times had certainly changed. This admiration was now directed at the fast-growing and aesthetically pleasing eucalyptus, its arcadian brio rousing enthusiasm from diverse profit-seeking parties within the real estate industry. By the early 1920s this newly reforested landscape containing vast eucalyptus groves and other nonnative species began to fulfill its intended development potential. Fast-maturing trees, organized in crowded groves and along curvilinear roadways, ushered in a number of new housing subdivisions.

As eucalyptus and other nonnative species were introduced to the area, a peculiar set of events and interactions unfolded. On the one hand, the physical environment of the East Bay hills reacted to the actions of developers and property owners who radically changed the natural setting; this occurred in the form of new growth patterns and the emergence of novel ecological dynamics. On the other hand, those natural systems—the aesthetics, smells, and feel of these new landscape attributes—also changed the perspective and behavior of local stakeholders and landowners as new management practices, environmental values, and development patterns emerged. Here we see an example of how human and nonhuman entities coproduce one another. Eucalypts for example may be viewed both as agents of *effect* introduced to fulfill particular economic interests through the construction of *desirable* neighborhoods, and as agents of *affect* capable of being *desired* through a process that cultivates aesthetic and emotional connections with surrounding community members. It is through this interactive process of social-ecological change that both risk and wealth increased on the landscape.

By the early 1900s the north hills region of Oakland was transitioning into "a residential area best known for its spectacular views, forested character, winding streets, and hillside architecture," a transformation marked by its eventual evolution from "a lumbering center" into a "vacation retreat for San Franciscans."[17] In 1923 for example the Oakland Hills witnessed a 900 percent increase in home construction over the previous five years.[18] If redwoods provided an important raw material for constructing a burgeoning San Francisco Bay Area during the 1850s, eucalyptus, Monterey pine, and other nonnative plantings during the 1920s provided the material base enabling the construction of a suburban respite from the hustle-bustle of San Francisco's now frenzied business environment.

Utilizing the areas' new scenic tree cover and countryside atmosphere, Oakland Hills property owners leveraged the very wealth their hillside timber resources had helped create, to reacquire capital investments from

San Francisco elites. The early 1900s saw the Oakland Hills area become a newly prized location for primary and second home acquisitions by Bay Area residents and new arrivals alike. As one real estate booster put it, "These [San Francisco] new-comers . . . found themselves as far removed from the dirt and turmoil of the work-a-day world as if they had traveled fifty miles into the mountains."[19] This transformation from an area that exported natural resource capital *to* the San Francisco Bay Area into a community that actively received financial investments *from* Bay Area communities has contributed to increased fire exposure and risk every step of the way. The East Bay hills of Oakland are therefore an early elucidatory example of broader, contemporary patterns of profitable suburban investment and expansion unfolding around the West.

HOME CONSTRUCTION AND THE CREATION OF RISKY REAL ESTATE

Three factors—timber harvesting, afforestation activities, and home construction—influenced the formation of elevated fire risk in hillside areas. The first of these activities introduced crucial infrastructure and physical capital, which in turn facilitated a second process: an extensive home development syndicate promoting reforestation in areas historically susceptible to fire. This newly forested landscape led to a third set of activities, as infrastructure originally laid to support forestry and grazing activities became the physical pathway for subsequent housing construction efforts. This marks the recursive, iterative, and mutually reinforcing nature of wealth and vulnerability.

By World War II the Oakland Hills landscape contained thousands of homes amid fast-growing vegetation cover, nestled within steep canyons and along protruding ridgelines. According to various government reports the proliferation of eucalyptus trees in particular has had the long-term effect of heightening fire risk and levels of social vulnerability in the area. Five years prior to the Tunnel Fire *The North Oakland Hill Area Specific Plan* mentioned that "in addition to naturally-occurring fires, the potential for accidental fires has increased as a result of . . . plant species such as eucalyptus and highly flammable ornamental vegetation."[20] It therefore comes without surprise that a Federal Emergency Management Agency (FEMA) report issued after the Tunnel Fire declared that eucalyptus and Monterey pine are "highly vulnerable to rapid fire spread" because they "release massive amounts of thermal energy when the burn. They also create flying brands, which are easily carried by the wind to start new spot fires ahead of the fire front."[21] Decades of fire suppression in the area only exacerbated this accumulation of flammable biomass, evidenced by dense tree stands, thick understory vegetation, and a buildup of leaf, branch, and bark litter.

The area ultimately seared by the Tunnel Fire experienced increased fire risk over many decades as part of a process of wealth accumulation that directly benefited real estate developers and property owners around the Bay Area. After the first commercial round of timber, residential fire risk gained traction, slowly being inscribed into the hillslopes through road grading and rudimentary infrastructure construction. Fifty years later home developments were constructed along these and other access corridors in areas that contain historically high levels of wildfire activity, as Map 3 in Chapter 1 depicts. Vulnerability to fire within these communities was further augmented through the introduction and maturation of a property value–enhancing replacement tree cover composed to a large extent of flammable eucalyptus species and Monterey pine.

Oftentimes discussions like this, which focus on the effects of vegetation change on elevated levels of composite fire risk, tend to overlook another important flammable feature on the landscape: the homes themselves. Perhaps the consummate social artifact, homes—created for human occupation, by human ingenuity, and with human articles—contribute to net vulnerability by adding substantially to a region's overall fuel load. Additionally most homes are built from and filled with wood and petroleum-based materials, making them comparable in composition to compressed Duraflame logs, which are of course manufactured and sold precisely because of their combustibility. The placement of these combustion boxes on the landscape in high numbers and in close proximity to one another was certainly the case in the Oakland Hills. According to an East Bay Regional Park District's report, "Many structures that exist within the interface are wood-framed or have wood shingles further increasing the complexity of wildfire risks. . . . Homes generally present fires with densities of flammable materials that are much higher than the surrounding wildlands."[22] But most importantly these homes increase vulnerability to fire because families—ranging from grandparents to infants—occupy them. Without homes there would be far fewer people. And with considerably less people social vulnerability to fire would be negligible. When viewed in this context, the contribution of eucalypts to increased fire risk has less to do with their material flammability and more to do with their ability to entice home construction activities.

By the beginning of World War II the Tunnel Fire area and its surrounding hillsides stood as a region notable for its high fire risk, flammable landscape, and vulnerable residential population. This chapter (indeed this entire book) has suggested that simply noting a population *is* vulnerable misses a much more interesting, albeit complex question; namely *how* it became vulnerable in the first place. As prominent vulnerability studies scholar Neil Adger aptly notes, vulnerability "does not exist in isolation from the wider political economy of resource use."[23] Deforestation and afforestation trends in the Oakland Hills were instrumental in growing the Bay Area economy, as capital flowing west to San Francisco eventually turned course, delivering substantial financial payments through land and housing development investment channels back across the San Francisco Bay into Oakland. Vulnerability to the Oakland Hills fire and other historical East Bay fire events can thus be productively viewed as the by-product of deeply intertwined regional economic development and resource use histories. These are the kinds of complex—and difficult to pinpoint—financial incentives, planning practices, and resource management decisions that drive the Incendiary around the entire American West. From the Pacific Coast Ranges to the Rocky Mountains contemporary settlements in high fire activity areas around the West, and the considerable costs and risks that accompany them, are nearly always linked back to historical and profitable land and resource use practices. Chapter 4 carries our analysis forward by assessing the affluence-vulnerability interface in the Tunnel Fire area from World War II to the fire event itself.

4 Setting the Stage for Disaster

REVENUE MAXIMIZATION, WEALTH PROTECTION,
AND ITS DISCONTENTS

By the end of World War II the Oakland Hills looked markedly different from the open grassy slopes and dense pockets of hardwood forest confronting European settlers a century prior. The hillside now comprised new, widespread, and arguably denser and more flammable tree cover than before. The intentionally designed arcadian environment was crisscrossed from foothill to sweeping vista by an expansive and undulating network of narrow, winding roads. And perhaps most importantly the whole region was populated with thousands of one-, two-, and three-story residential homes placed precariously within steep ravines and atop steep ridgelines.

SUBURBAN HOMEOWNER POLITICAL MOVEMENT AND CURTAILED CITY REVENUES

Several decades after these housing development and afforestation activities first began, conditions of vulnerability in the Oakland Hills were deepened by a second powerful force: increased suburbanization and extensive changes to preexisting tax revenue structures in the state of

California. Building momentum during the 1950s–1960s, a powerful suburban homeowner political movement reached a groundswell in the late 1970s resulting in dramatic changes to California's taxation and revenue collection system. These changes were in part a by-product of skyrocketing real estate values and surging inflation rates that resulted in elevated private property taxes. Members of an emerging conservative antitax (or "tax revolt") movement launched a massive publicity campaign attacking these fast-rising property tax payments. To rally support the movement focused much of its criticism on public expenditures in older city segments with high African American, Hispanic, and Asian American populations far removed from the largely white suburban homeowner tax base. The argument was simple: in an era of soaring home ownership costs why should hard-working, responsible, and self-sufficient California citizens have to make exorbitant tax payments to support others who have become dependent upon costly government programs?

Public sentiment across California was put to the vote in 1978 in the form of Proposition 13. The tax revolt initiative won overwhelmingly (64.8% to 35.2%), leading to the passage of the nation's first comprehensive tax limitation measure. Voting patterns were not the same across all cities however. In Alameda County for example suburban cities approved the measure with more than 70 percent of the vote, while 52 percent of the residents in Oakland rejected the measure (see Map 5). This voting pattern underscores an emergent postwar metropolitan development sentiment around the state and nation: a desire by fast-growing, white, suburban populations to substantially detach their tax payments, investments in municipal programs, and general wealth from core urban areas.

The effects of Proposition 13 were nothing short of profound for California's city economies. Real estate property taxes were reduced by reassessing properties to their 1975 value. Furthermore the proposition set maximum tax rates at 1 percent of total property value and restricted maximum increases in assessed value to 2 percent from one year to the next. Of even greater consequence its passage mandated that property could only be revalued under a transfer of ownership. The impact on city revenues was particularly significant for cities like Oakland with aging infrastructure, a large working-class population, and sizable public works programs.

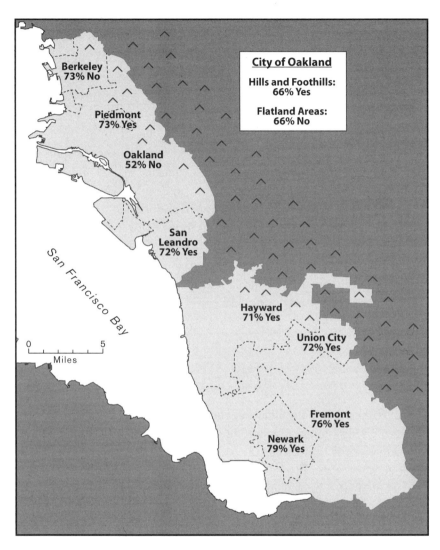

Map 5. California Proposition 13 voting outcomes in Oakland and its suburban neighbors in 1978. (Adapted from Self, 2003, by Peter Anthamatten)

In order to fully appreciate the depth of expenditure curtailments in cities like Oakland, it is important to contextualize Proposition 13 and the tax revolt within broader patterns of postwar suburbanization in California and around the American West.[1] During this period numerous suburban communities received substantial levels of public and private investments to attract homeowners, industry, and financial capital—marking a redistribution of wealth, land value, and tax revenues that directly undercut the economic best interests of core cities like Oakland. This process was spurred by large government subsidies in the suburban housing market and a growing coalition of construction entrepreneurs and suburban city boosters. Through federal urban policy retrenchment and market forces the suburban landscape was converted into various forms of capital: increased property values for homeowners, direct profits for developers, and taxes for public agencies.[2] The results of regional suburbanization became apparent in the rapid development of suburbia and the under- (or at least stagnant) development of Oakland during subsequent postwar decades.

REDUCED FIRE DEPARTMENT BUDGETS AND VULNERABILITY INTENSIFIED

Urban residents in cities like Oakland are now paying for many services through price systems composed of fees and charges rather than general citywide revenue sources like the property tax. And yet despite these new revenue sources the adverse funding consequences for the Oakland Fire Department are unmistakable. Prior to Proposition 13, revenues from city property taxes were almost entirely earmarked for city expenditure items. This is a legacy of California's "Home Rule Power" and "Separation of Sources Act" established in the early 1900s. Together these stipulations granted cities power to draft their own charters, govern municipal affairs, and retain locally generated taxes.[3] For example at the time of Proposition 13's passage, 90 percent of fire department budgets in California were funded through local property tax revenues.[4] After the measure's passage local property tax revenues witnessed a 53 percent overall reduction.[5] By the 1990s the proportion of city revenues drawn from property taxes had decreased to 8 percent within the state. This decline is evident at the county level as well. Prior to Proposition 13 counties in California drew 33 percent of their revenue from property taxes; by the 1990s property taxes shrank to a mere 10 percent of county revenues.[6]

Sharply reduced city tax revenues resulted in underfunded fire response, mitigation, and retrofitting programs throughout Oakland.[7] According to FEMA the Oakland Hills were "particularly vulnerable in the fall of 1991, after 5 years of drought, several months with no recorded precipitation, and reduced efforts to control wildland interface fires due to State and local budget limitations."[8] The report goes on: "Before budget reductions in the 1970s and 80s, [the Oakland Fire Department] was recognized as one of the strongest fire suppression departments in the western United States. The budget limitations reduced the number of companies in service and the staffing on each company. Several stations were closed during this period."[9]

Moreover it is well documented that damaged and dead trees pose a significant fire risk. A deep freeze in 1972 and again in 1990 harmed numerous eucalyptus groves and contributed to the formation of dry and highly flammable vegetation cover and ground litter. Yet according to the Oakland General Plan, budget reductions led to conditions where "the City lacks the funds to completely restore its damaged or dead vegetation."[10] Still further, according to the fire chief, Oakland's Fire Department operational units were overextended during the 1991 Tunnel Fire, leading to diminished emergency response and mitigation capabilities.[11]

Meanwhile many understaffed firefighting agencies in the West are now facing extended firefighting seasons as warmer weather, less precipitation (particularly as snow), and increased aridity stretch into winter months. Long periods of winter red-flag fire days are no longer abnormal. Even Cal Fire director Ken Pimlott mentioned that experiencing midsummer fire conditions in January "really is unprecedented. In my career, I've not seen this level . . . this is what we're up against." This quote was all the more significant given he was referencing a fire that had just burned thirty-three acres in four days in Humboldt County in far Northern California—a region usually sufficiently moistened with rain at that time of year.[12] For many units around portions the West, one fire season tends to bleed into the next. This trend means firefighting costs are being accrued throughout a greater portion of the year, leading to higher demand for resources during a period of significant public fiscal constraints around California and the West more generally. The firefighting budget candle is indeed burning from both ends.

In light of budgetary constraints (and a greater burden being placed on firefighting agencies), many fire departments now rely on a regional network of response agencies. This model of flexible and shared emergency response governance—commonly referred to as "mutual aid"—is exceedingly more cost efficient for participating municipalities than fully staffed, autonomous units. And yet this interagency structure has its own shortcomings. According to a State of California report, during the Tunnel Fire, "nozzle hook-ups for Oakland ... actually had a smaller size than the other districts. So, when firefighters came from areas, they could not plug into the Oakland hydrants." While some have suggested that appropriate hook-ups were available along lower portions of hydrants, the fact that coordinating agencies and participating firefighters did not know about these modulations again speaks to a lack of funding—in this case for mutual aid education and interoperability training.

The account goes on: "Communications broke down, they could not communicate with one another adequately because the radios were on different wave lengths. . . . It was pretty much chaos."[13] The vital Incident Command System (ICS), which provides a reliable authority structure where real-time information flows up and response commands flow down, was also compromised by a lack of consistent communication protocols and technology applications. By early afternoon of the Tunnel Fire various nearby departments were unable to adequately participate in the ICS system. This hampered interagency coordination and increased confusion between mutual aid participants.[14] Along with technical and communication incompatibilities, many of these agencies were themselves undergoing similar operational constraints from budget cuts.[15] The Tunnel Fire illustrates how residents in the city of Oakland—who had become increasingly dependent on "flexible" and "efficient" cross-jurisdictional support—were underserved by this mutual aid structure.

This form of service provisioning, based on cost-cutting and efficiency-seeking imperatives, presents the limitations of neoliberal urban policy. A complex process of suburbanization—characterized by the reorientation of public and private finance toward metropolitan peripheries and the emergent influence of a homeowner antitax movement—contributed to budget cuts, decreased city revenues, and ultimately the reduction of fire prevention and mitigation capabilities by the very units (cities and municipalities) traditionally responsible for those activities. Moreover the regional mutual aid fire response model emerging since budget reductions was itself rendered vulnerable to technical and communication interoperability limitations brought about by the hasty integration of diverse agencies. Additionally, according to one Bay Area firefighting professional, mutual aid also leads to the presence of responders unfamiliar with local landscapes, evacuation routes, and hazard behavior. In sum, as the cost of fire mitigation decreased under this neoliberal model, so too did the quality of those mitigation efforts.

CITY REVENUE EQUALIZATION AND NEW HOUSING SUBDIVISIONS

The effects of city revenue curtailment on vulnerability extend beyond reduced fire prevention and mitigation capabilities. Beginning in the late 1950s city builders in Oakland erected hundreds of high-density housing units in areas directly adjacent to the origin of the Tunnel Fire (see Figure 2 in Chapter 1). The Hiller Highlands Complex, Parkwoods Apartments (see Figure 13), and other new developments in Oakland's hillside areas were constructed in part as a response to lost revenues from state and federal retrenchment and state property tax restructuring. Home developments and subdivisions can generate substantial city income, particularly in hilly terrain containing high property values and tax revenue potential.[16] Table 1, based on comprehensive city tax roll analysis, details the considerable property tax revenue generated from homes in state-designated high fire risk areas of Oakland during 2012.[17] This single-year analysis is illustrative of the considerable revenues that are produced in other years and within suburban landscapes around the American West. As part of its strategy to offset decreased property tax revenue growth rates, the City of Oakland extended housing developments higher and higher into the hillside.[18] Unfortunately,

Table 1 Comprehensive city tax roll analysis from fiscal year 2012 reveals that very high fire-risk areas of Oakland account for a significant portion (nearly one-third) of the city's overall property tax revenue. Property taxes from these high-risk areas generate more revenue per unit (57% higher) compared to the rest of Oakland. (Credit: Alejandra Uribe)

	Property Tax Revenue (2013)	Area (sq mi)	Number of Units	Tax Revenue / Unit
City of Oakland	$334,655,556	56.07	69,749	$4,798
Very High Risk (VHR) Area	$109,667,332	17.32	16,491	$6,650
Non-VHR Area	$224,988,224	38.75	53,258	$4,224

but not unsurprisingly, many of these expansive home and condominium developments, located at the Tunnel Fire epicenter, were the first units to be destroyed in the 1991 blaze.

With a vast swath of hillside homes—many of which are in high-density housing developments—Oakland has proven its ability to develop, maintain, and extract profit from this flammable yet lucrative hillside. *The North Oakland Hill Area Specific Plan* describes and reaffirms this revenue replacement strategy: "Given the assumed value of new homes in the North Oakland Hills Area, and the significant level of property taxes generated, the net fiscal impact of development on public services is positive." The plan goes on to note that this net fiscal benefit to the city holds true "unless an attempt is made to operate a new fire station."[19] Here city officials seem to indicate that tax revenues from new housing developments in the Oakland Hills can have a positive fiscal impact on the city but that net revenues will only increase *without* the construction and maintenance of a new fire station to serve the area.

In a disturbing revelation this means that populating the hillslope to compensate for lost revenues not only requires intentionally placing homes within a landscape historically exposed to frequent wildfires; it requires doing so without additional fire protection. This despite city planners soberly noting in the same report, "New residential development will significantly increase the potential for loss of life and damage to property from fire hazards in the North Oakland Hills, especially given the

Figure 13. A high-density condominium complex that burned during the Tunnel Fire is rebuilt in its original location, nestled into open space at the city's edge. The 1991 fire's origin is located just upslope. The fire swept quickly through the complex before jumping ten lanes of highway and spreading south, consuming thousands of homes along the way.

poor accessibility." The ripple effect of reduced tax revenue potential in Oakland is clear: the construction of *more* housing units with *less* fire protection in high fire risk areas.[20]

The Ongoing Debate over the Financial Solvency of New Housing Developments

The debate over new housing developments in designated high fire risk areas is seemingly intractable. From a wildfire prevention standpoint those in opposition to new developments often cite the increased cost fire mitigation activities will add to already unwieldy city budgets. Newly developed hillsides not only expand firefighter response areas; they also increase the total area necessitating ongoing city mitigation and monitoring activities. These are costly activities that increase the workload of an already overextended fleet of fire and emergency response agencies. As a land manager for the East Bay Regional Park District (EBRPD) put it, "There are a lot of proposals to put houses on steep, steep ridges surrounded by dense woodland. They want us to provide fire protection and we're not going to be able to. And the people moving into these areas won't know it. It's asking for trouble."[21]

Unsurprisingly, local development interests hold a very different perspective. Builders and progrowth advocates in East Bay hill areas, for example, remained undeterred even in the months and years after the Tunnel Fire. For these individuals new home construction—particularly high-end homes in valued hill locations—promised to deliver new property tax revenues for the city and a consumer base capable of attracting merchants and boosting local sales taxes. As one owner of a large developable area in the city of Pleasanton put it, "What we have seen and continue to see is that all the mitigations you want, to prevent such a disaster, are being built in (to new projects)."[22] In fact to many, including a onetime prodevelopment mayoral candidate in Union City, California, the 1991 firestorm only strengthened the case for new residential communities by ushering in tough new fire prevention ordinances such as strict home material ordinances, vegetation clearance, setback rules, and indoor sprinkler requirements. Moreover broader infrastructure management lessons were learned from the Tunnel Fire, such as the need to have clearer and more accessible evacuation roads and backup power generation resources for pumps sending water to crucial upslope reservoirs. These are lessons that would shape the development of new—and advocates would argue more fire-adapted—residential communities.

While each side of the debate can make compelling arguments to support its position, a truism emerges in nearly every case: meaningful tax benefits from these developments only materialize if long-term infrastructure operating budgets are kept to a minimum. Expanding the city property tax base was certainly a strategy employed by the City of Oakland during the 1950s–1970s. But this was only a net positive fiscal endeavor for the city because very little was spent to expand preexisting firefighting services. Indeed the city has become increasingly dependent on mutual aid support comprising the likes of Cal Fire, EBRPD, and other members of the Hills Emergency Forum. As a homeowner association president in Oakland noted in response to another nearby proposed development, "They (project supporters) keep saying the city will get half a million in tax revenue from this project (each year). But that's without the cost of opening the fire station." Ultimately it will always behoove prodevelopment interests, including landowners and builders, to tout the fiscal advantages of residential development, as they have nothing to lose and everything to gain. However, when a synoptic view is taken, we see that real financial incentives for cities only hold in the presence of reduced (or at least fixed) fire services. And although fixed services may be offset by new construction regulations, large firestorms (like the Tunnel Fire) do not conform to new design features such as indoor sprinkler systems or modified eaves.

As the story of California tax restructuring and local development responses indicates, the forces behind the Incendiary shape-shifted in the Tunnel Fire area in the decades after World War II. The expansion of homes and increased fire risk was driven by the complex relationship between government retrenchment due to the reorientation of wealth and investment into suburban landscapes; neopopulist homeowner policies predicated on individual rights, estate-based wealth protection, and a nearsighted commitment to social responsibility; and resulting tax revenue losses and depleted operating budgets within tax-dependent city fire services. As is the case in most areas of the West, the actors, institutions, and processes composing the AVI—and driving the Incendiary—are complex, historical, multiscale, and profit driven in nature.

THE POLITICS OF UNEVEN DEVELOPMENT: RACE, CLASS, AND VULNERABILITY ENTANGLEMENTS

Delving deeper into the formation of affluent communities also presents a window through which we can explore its opposite: the systemic root causes driving the concomitant production of risk and precariousness for less privileged segments of society. Thus far *Flame and Fortune* has viewed vulnerability as a series of hillside (in Oakland) and suburban (across the West) manifestations, characterized by enhanced fire risk due to a rapid increase in total fuel load, elevated residential community hazard exposure, and impaired firefighting capabilities. While these landscape changes are extremely important, particularly for those residing in very high risk areas, there exist other important and interrelated vulnerabilities throughout the rest of Oakland (and other core metropolitan areas). Factors generating vulnerability and affluence in the Tunnel Fire area also contribute to the production of vulnerabilities within the flatlands of Oakland. Here the allocation of estate-based wealth for property holders and levels of net vulnerability are highly uneven across space and demographic groups—yet also deeply intertwined in their coproduction.

At first glance Proposition 13 voting patterns may appear to signal a tax revolt movement largely driven by suburban city residents and rural homeowners. However as one digs deeper into the spatial distribution of Proposition 13 voting in Oakland (Map 5), another trend emerges: voters in the Oakland Hills followed suburban patterns in support of the proposition by a nearly 2 to 1 margin, while downtown Oakland and flatland areas, composed much more heavily of working-class minorities, voted 2 to 1 against the measure.[23] This intracity division in voting illustrates how residents in Oakland hillside and flatland areas clearly held different views on what constitutes acceptable property tax rates and levels of income sharing. This voting outcome reflects several key demographic and socioeconomic patterns and differences across the city. For example, flatland areas contain a much higher number of property renters who did not stand to benefit as directly from the passage of Proposition 13. These residents were also generally less affluent than hill and suburban populations and so utilized—and thus valued—local tax-supported services to a much greater extent.

From the perspective of many upslope and suburban residents the spatial distribution of votes in favor of Proposition 13 underscores a powerful antitax sentiment supporting the protection of personal estate-based wealth over redistributive structures of taxation. However, as the previous section illustrated, the cumulative impacts of city revenue curtailments on hillside residents had unintended consequences: the rapid depletion of fire department budgets and subsequent housing construction in fire prone-areas. In the effort to reduce individual property taxes, fire response agencies were left with insufficient resources to protect those very same vocal property owners from devastating wildfires.

The case of the Tunnel Fire suggests that the acceptance of high fire danger and curtailed fire mitigation capabilities by these hillside residents is buoyed by a multilevel system of fire risk subsidization.[24] This structure of risk-offsetting is supported to a large extent by an expansive insurance industry supporting loss indemnification. Moreover beyond the insurance industry stands a large federal system of fire mitigation services that reduces insurance company cost-share responsibilities, leading in turn to artificially low coverage plans for homeowners. If residents can afford the cost of comprehensive fire insurance—such as guaranteed replacement cost plans—they can effectively pay for the right to live in areas with historically high fire activity.

The relatively low cost of insurance and resulting levels of risk-security for homeowners is influenced in part by a tiered structure of postdisaster redevelopment policies. Immediately after the Tunnel Fire for example the State of California paid an estimated $15 million to local governments in the form of public disaster assistance. This included payments issued directly to city governments, loans to owner-occupied and rental properties, individual and family grants, and homeowner property tax deferrals. Meanwhile federal grants were issued for an estimated $42 million to state and local governments to recover these and other incurred costs.[25] The influence of these pre- and postdisaster policies on levels of vulnerability is significant because they help Oakland Hills residents rationalize and tolerate (as expressed through Proposition 13 voting patterns) chronic citywide disinvestments in fire response and prevention programs.[26] With federal, state, and private resources available to facilitate risk reduction, the implications of reduced local firefighting budgets become less acute and easier to accept.

To be sure, residing in these hillside areas presents very real risks for community members—including loss of life, bodily and psychological trauma, and damage to or loss of irreplaceable items, heirlooms, and keepsakes. And yet despite these risk-offsetting policies many hillside residents living in fire-prone areas hold a lower level of *net* vulnerability when compared with less privileged minority residents in flatland areas.[27] Poorer city residents who cannot afford comprehensive insurance premiums and are thus rendered insufficiently indemnified may therefore feel the acute and uneven social consequences of elevated fire risk. While homeownership requires a base level of insurance coverage, these may not be full cost recovery plans; and renters in Oakland and other areas of the West are even more susceptible to loss from building fires when rental insurance plans are too expensive. In light of these conditions depleted local firefighting capacities result in a landscape comprising individuals with highly differentiated abilities to recover from and cope with natural hazards. Post–World War II antitax homeowner politics coupled with available insurance and government risk-reduction measures illustrate why different members of Oakland will experience fires (even single-structure fires) differently.

Of course decreased firefighting capabilities represent but one form of social precariousness resulting from city revenue curtailment. Reduced welfare and financial hardships are also generated from declining support for other nonfire-oriented public programs, such as those related to education, environmental health, psychological services, and child care. For less privileged residents in the flatlands of Oakland the impacts of city revenue curtailments run deep, as many private schooling options, toxic cleanup practices, and other expensive municipal resources remain fiscally out of reach. Moreover the impacts of disinvestment cut across generational lines (i.e., those purchasing homes pre– and post–Proposition 13) and thus disproportionately affect new and immigrant households who must buy into newly reassessed properties with adjusted (and higher) property tax rates.[28]

Still further, tax restructuring has led city officials to pursue alternate income streams, for example by elevating sales tax rates. Sales taxes however disproportionately impact lower-income families who spend a larger portion of their salary on household staples.[29] Similarly state and local budgets have become more reliant on income taxes for revenue. Yet income tax revenues are among the most volatile forms of state funding, as government entitlement programs experience greater susceptibility to budget crises during economic downturns and periods of high unemployment.[30]

Because local property taxes support city fire services (even despite post–Proposition 13 revenue curtailments), flatland residents wind up paying for fire prevention and mitigation programs in high-risk areas despite not experiencing their direct benefits. What has emerged then is an inverted form of the antitax argument. City residents in one part of the city are expected to pay property taxes that deliver revenues supporting city expenditures in other parts of the city. A key difference of course is that flatland taxpayers are generally less well-off and in a more tenuous position financially compared to hill residents. Placing a considerable portion of the fire mitigation cost burden on impoverished communities that do not benefit from the bulk of those mitigation services further highlights the relationship between affluent hillside developments and the production (and intensification) of flatland vulnerabilities.

In short, flatland residents experience elevated vulnerability to fires as a result of decreased fire prevention and response services, and also because many households lack the capacity to recover losses due to unaffordable comprehensive insurance plans. Adding to the burden are other potential acute impacts of city revenue curtailment where the responsibility of balancing city budgets is shifted onto household income, expenditure, and property tax activities. A wider analysis of Oakland thus reveals the uneven distribution of risk and vulnerability as it unfolds across time and space. Factors producing fire risk in the Oakland hills (like other contemporary suburban areas)—which are influenced by an underlying suburban homeowner mentality premised on estate-based wealth protection, antitax sentiments, and a nearsighted commitment to revenue sharing—are closely linked to the formation of other urban vulnerabilities across the city.

"A Part of the City Below"?

After nearly 150 years of development in Oakland, perhaps the most acute burden of risk from reduced city spending (including risks from urban fires) besets flatland community members receiving only attenuated benefits from the region's history of instrumentalist land use policy, lucrative real estate developments, and skyrocketing property values. Map 6 shows how the financial dividends of homeownership in Oakland are disproportionately higher in hill areas and in particular the area impacted by the 1991 Tunnel Fire. Here analysis of seventy years of census data reveals that median fire-area property values increased from under $100,000 in 1940 to about $900,000 (a 900% rise in value) in 2010, while increases in

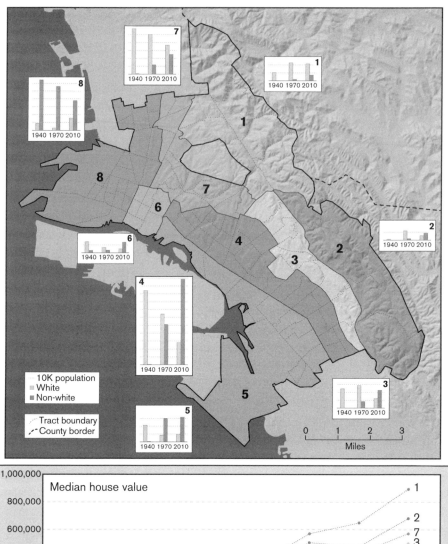
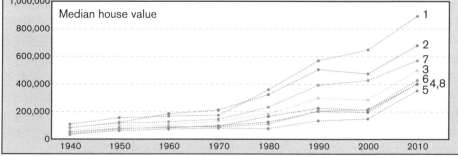

Map 6. Changes in home values from 1940 to 2010 illustrate the differentiated allocation of estate-based financial benefits across space and demographic groups in Oakland. (Credit: Peter Anthamatten and Alejandra Uribe)

flatland areas grew more modestly from nearly $70,000 to roughly $400,000 (a 570% rise in value) over the same period. Moreover the most dramatic increases in home value have occurred in areas composed predominantly by white residents, while segments of the city with higher nonwhite populations have experienced considerably lower increases in home value. This bifurcated development reflects historically differentiated capital investment and land use planning in Oakland's flatlands through a process referred to as "demarcated devaluation."[31] In other words these are not accidental or unexpected outcomes but rather the legacy of historical land use planning decisions developed intentionally to foster high property value increases in some areas and less growth in others.

To some it might appear that the sentiments offered in an early promotional real estate brochure still hold true: Oakland Hills residents are "a part of the city below, yet apart from it."[32] That marks a convenient, sanitized, and apolitical depiction of the city and its history. Quite the contrary, dedicated relational and historical analysis reveals that these bifurcated development trajectories and levels of vulnerability are in fact deeply intertwined. These areas share much more than just a city boundary. Wealthy, white hillside communities and relatively poorer, minority flatland residents are indeed a part of the *same* urban story. Turn-of-the-century sylvan neighborhoods were generated through the introduction of property-enhancing replacement tree cover and other amenities leading to the development of a "Mecca toward which the successful Californian turns."[33] These land use developments, as economic geographer and theorist David Harvey famously notes, represent "manifestations and instanciations... of... particular social relations."[34]

And it is through this process that we are able to identify the formation and ossification of social stratification—local differences reflecting what the geographer Dick Walker describes as "unevenness within larger geographies of capitalist development, industrialization and urbanization."[35] The city of Oakland matured in a manner perhaps most notable for its spatially uneven development where both space and nature were continually produced by and for estate-based capital accumulation. Indeed the settlement of hillside landscapes by wealthy, white communities was no accident. As a 1911 Laymance Real Estate Company brochure stated about one neighborhood in the Oakland Hills, "It is probably unnecessary to even mention that no one of African or Mongolian descent will ever be allowed to own a lot in Rockridge or even rent any house that may be built there."[36]

Some have suggested that Proposition 13 voting patterns such as those found in Oakland reflect dramatic differences in real estate values, where support for Proposition 13 simply increased in tandem with anticipated tax savings.[37] *Flame and Fortune* does not refute this argument. Instead it has proposed expanding upon its rather aspatial, ahistorical, and apolitical explanatory purview by connecting early land transformations to these voting patterns a century later. Beginning in the mid-1800s resource extraction and subsequent real estate speculation activities contributed to the production of white, affluent communities in the hill area. Over a century later Proposition 13 dismantled structures of revenue redistribution in a manner that conformed to social stratification and real estate value disparities across space. Voting patterns wound up reinforcing these divergent development trajectories first established under early land speculation activities. This process illustrates *why* disparate land values exist and how they *contribute to and reinforce* (in part through subsequent voting patterns) levels of vulnerability and fiscal precariousness in flatland areas due to chronic city disinvestments.

This part of the book illustrates how after nearly 150 years of urban development perhaps the greatest risk burden from city disinvestment besets flatland residents—a suite of communities receiving only attenuated benefits from the region's long history of lucrative land use and real estate developments. Beginning with early land transformations described in Chapter 3, Oakland's development highlights how the production and reproduction of difference matters, and hence why it is important to link the production of wealth and vulnerability over time *and* space. In this way lucrative hillside real estate developments and postwar tax revolt movements are inseparable from, indeed constitutive of, the lives and livelihoods of inner-city residents. Here we see the dynamic and political ecological nature of vulnerability—where the allocation of estate-based capital accumulation and levels of net vulnerability are highly uneven yet also deeply intertwined and coproduced.

5 Who's Vulnerable?

THE POLITICS OF IDENTIFYING,
EXPERIENCING, AND REDUCING RISK

As turbulent morning transitioned to tragic afternoon, flames and swirling winds ignited one home after the next, engulfing whole blocks within minutes. Engine Company 19 had been fighting spot fires when overwhelmed members of the team took refuge near Gwin Tank just north of the fire's origin. Team members attempted to leave their position on several occasions but were repeatedly turned back by flames and low visibility. Stuck in defensive position, Engine 19 issued an urgent report to the command unit.

"Command, Engine Nineteen."

"Go ahead, Nineteen."

"Yeah, Tunnel Command. We are trapped behind a tank. We've abandoned our rig. We need help to get out."

"Fifteen is right above you. Fifteen, did you copy that?"

A few moments later Command continued,

"Whatever's got you trapped, if you keep a big line off that hydrant you'll protect yourself."

"We'll protect ourselves, Command . . . But don't forget us."

Engine 19 delivered these three words—"Don't forget us"—with a collected and determined tone, as if to emote valor and a sense of control over the increasingly dire situation. Yet a closer listen also reveals a less confident timbre near the end of the short passage in a slight rise of pitch—a subtle distress signal and appeal for help. It is as though the firefighter does not want to reveal his fear yet also recognizes this moment as perhaps his last opportunity to convey the seriousness of his engine's predicament. Moreover (and tone aside) the passage itself is performative. It *is* a plea for help. It emanates from an individual—forced to take refuge behind a water tank and impinged by flames—who has been laid bare and stripped of his bravado, ego, and unassailability. "Don't forget us" is an admission of vulnerability. It is a brief yet powerful reminder of the visceral, deeply profound, and frequently disguised experiences of embodied vulnerability that afflict emergency responders during disaster events. (The firefighter in question was eventually able to leave the area.)

These subtle and oftentimes veiled expressions of exposure, fear, and trauma arose throughout the Tunnel Fire area, and not just from firefighters. Area residents also embodied real risks, trepidations, and anxieties—many of which would not be readily apparent to the untrained eye. Just as a cursory observation of firefighters may lead one to believe they are too brave to feel vulnerable, a hasty evaluation of households in the relatively affluent Tunnel Fire may lead some to question their capacity to experience substantive vulnerability. However, reading between the lines and identifying the subtle (and not so subtle) "Don't forget us" moments for each resident reveals many difficult circumstances as well as "the psychological and emotional maze," as one resident put it, impacting affected households. And yet as later sections of the chapter also elucidate, these risks and vulnerabilities may be partially offset by the ability of affluent residents to leverage their privilege and secure crucial resources and advantages for the entire neighborhood. This chapter therefore aims to shed light on the diverse ways loss and trauma are experienced and abated in affluent, suburban communities.

VULNERABILITY AMID AFFLUENCE: EXPLORING THE CONNECTIONS

Rather than use vulnerability to explain fire-prone landscapes, it is vulnerability itself that deserves further explanation. What does it actually mean to be vulnerable to wildfires, particularly in urban peripheries comprising communities whose overall affluence and privilege are higher than in many other urban neighborhoods? These living standards lead many to take a less than sympathetic view of residents living in fire-prone areas—a perspective for example that partly influenced the urban historian and theorist Mike Davis to posit "the case for letting Malibu burn." For some critics many suburban residents appear to take on a level of *assumed* risk. This means that homeowners know fire hazards are a distinct possibility when moving into areas such as the Oakland Hills. This stands in contrast to *imposed* risks, which involve the emergence of unexpected threats occurring without prior knowledge of local residents—risks that may for example derive from sudden planning decisions that drastically alter the rate or severity of hazard exposure for community members.[1]

This chapter seeks to complicate notions of vulnerability and reject the crude dismissal of risk within affluent communities. On the one hand, rather than oversimplify a diverse residential base and essentialize its precariousness as a direct function of geographic location ("They can't be vulnerable, because they live in a wealthy neighborhood"), this chapter instead argues for the presence of *variegated vulnerabilities* comprised in a landscape of residents, each with unique resources, finances, psychologies, family histories, and risk exposures. On the other hand, the presence of affluence and privilege cannot be ignored or treated as simply a benign community attribute. A shift from synoptic to household-scale analysis highlights many of the ways community members collectively leverage their financial privileges to minimize or offset certain risks. In particular the presence of individuals with the energy and ability to access crucial city resources is shown to help reduce levels of risk and deliver benefits to the rest of their fellow community members.

Household-scale Analysis and the Benefits of Antiessentialist Inquiry

In the Tunnel Fire area each household's experience with risk and vulnerability was unique. This may seem an obvious point but it is one worth repeating, particularly in affluent communities where all residents are assumed by some to contain a level of privilege that insulates them from any substantive, meaningful, or acute form of vulnerability. I have given many presentations on this topic over the past several years and in several instances audience members have raised questions about the validity of the claim that hillside households are in fact "vulnerable." This perspective tends to revolve around three assertions. First, these are households with considerable resources (such as premium insurance plans) to help overcome and recover from such trials. Second, these residents knew there were fire risks when purchasing their property; so isn't it fair to say they assumed this risk? And third, claims of hillside precariousness seem overstated when placed in the context of a world with acute water and food insecurities, race- and gender-based violence, religious persecutions, and pernicious economic injustices. Using the term *vulnerable* to describe residents in the Oakland Hills could be seen as watering down the term, rendering vulnerability an analytically blunt concept. As a political ecologist, I would argue forcefully against such dismissive epistemology. Fine-grained analysis that scales down to household and "local" levels—a framework of analysis implemented below—is designed to substantiate, challenge, or overturn explanations premised on more synoptic and "distant" analysis.

As it turns out, the skeptical sentiment held by many actually finds a quasi-sympathetic ear from fire victims themselves. (Indeed, there *are* distinct estate-based benefits that emerge after a fire.) Nearly all residents are quite aware of the unique ways in which loss and tragedy are manifest and experienced from one household to the next. Perhaps more crucially they question the way vulnerability is uncritically assessed, as if the whole firestorm and its aftermath hold a common tragic denominator. A married couple whose home burned in the fire commented on the challenge of applying the term *tragedy* to the firestorm.

> I do not view [the Fire] as a tragedy.... There is loss. My children's childhood artwork is lost forever. But tragedy is the professor I met who has pain walking because he was burned so badly and whose wife and dear friend died in the fire. That's tragedy, ok. Losing your children's artwork is painful, but it's not tragic.

The husband added,

> It interests me how there is so much hyperbole around the fire.... The number of people I heard say after the fire, "It looks like Hiroshima up there." Which is ridiculous. But we have a friend ... who walked through Hiroshima two weeks after the bomb.... And they lost their house in the fire too. And I said, "Some people ... who are walking through the neighborhood, who don't live up there, [say] that it looks like Hiroshima. And he said, "Oh no ... it's more like the Battle of Manila. Chimneys remained." [*Laughing*] I loved it. That's a man with experience talking. Oh goodness.

Levels of vulnerability within the Oakland Hills are certainly not, on a day-to-day level, equivalent to the types of risk experienced by many residents in the relatively poorer flatlands of Oakland, let alone other more impoverished, marginalized, war-ravaged, and precarious regions of the world. But to engage in such comparative risk assessment for the ultimate purpose of dismissing an entire population as simply "not vulnerable" (or similarly to label an entire population as equally afflicted by tragedy) is to rebuff good political ecological analysis and reject nuanced, data-driven, nondiscriminatory inquiry. Rather analysis should be premised on the notion that vulnerability comprises a set of complex and contingent factors that are heterogeneous across all households and social groups.

VARIEGATED VULNERABILITIES: A TOPOGRAPHY OF RISK, LOSS, AND TRAGEDY

In the Oakland Hills area, much like other fire zones around the West, diverse risks and vulnerabilities characterize and delineate the experiences of thousands of households. These manifestations of risk and vulnerability fit within a series of psychological, physical, and financial categories connoting the different ways residents experience anguish, discomfort, pain, anxiety, and loss. In-depth interviews with residents illuminate how broad-brush descriptions of risk (including those that characterize entire neighborhoods as bereft of such vulnerabilities) are inaccurate and insensitive to the plight of many.

Resident accounts of risk and vulnerability fall into two broad categories; nearly all described conditions that contributed to increased exposure and hardship. They also detailed factors that exacerbated already difficult circumstances. We may understand these as *variables of cause* insofar as they shaped and conditioned potential or future experiences of risk. These include attributes such as geographical location, level of mobility, number of dependents, and race and gender association. Others described actual experiences of loss as a result of the fire. Here residents focused on the consequences of being vulnerable. These *variables of effect* were diverse in nature and included issues such as property loss, physical ailments, and emotional stress from displacement. One of the most interesting yet tragic aspects of vulnerability in the fire area is the way many attributes of risk, injury, and trauma become linked up with future manifestations of suffering and distress. Several residents' descriptions highlight the ripple effect of interconnected fire-related stresses and losses, as vulnerabilities extended outward from the event itself, shifting forms and reemerging in new, complex, and sometimes unpredictable ways.

Diverse and Interconnected Vulnerabilities:
From Reduced Mobility and Emotional Fatigue
to Lost Lives and Lost Memorials

Twenty-five individuals lost their lives in the fire. For many of these victims (as well as hundreds of others who narrowly escaped) physical mobility was a significant contributing factor to individual risk and hazard sensitivity. This was particularly true for elderly community members. Of the civilian fire fatalities six out of twenty-three were over seventy-five years old, and over half (12 out of 23) were over the age of sixty-one. In several instances elderly residents had trouble evacuating the fire zone. I still hold vivid memories of meeting my friend in between our two homes as the fire quickly approached (see the Introduction). A senior neighbor approached us in need of our immediate assistance. She led us to her home where we helped lift her frail husband up the front steps and into their waiting car. While they were able to successfully avoid the oncoming fire, many others were not so lucky. Many senior and physically impaired residents initiated evacuation measures only after it was too late, and after most of the neighborhood had already departed. Unable to find assistance, they were left to fend for themselves.

In one example the burned remains of an infirmed senior citizen were found near her home's entryway. Tragically this elderly resident, who was home alone, was unable to flee the onrushing fire—she was presumably consumed by the flames while trying to crawl to safety. Two days after the fire, as neighborhood members finally got permission to visit the fire zone, the elderly woman's granddaughter was walking through the rubble when she came upon the body. This was a profoundly unsettling discovery on many levels. On the day of the fire the young girl had called 911 for help but was told by the operator that everyone had been evacuated. She accepted the response and ceased further efforts to find her grandmother. That moment of discovery would continue to weigh heavily on the granddaughter's conscience. She would struggle for many years with various mental health issues. Seventeen years after the fire her deceased body was found in the location where she had found her grandmother's remains. Although many other issues contributed to her suicide, family members have stated quite confidently that the events of the 1991 fire significantly influenced the young woman's tragic demise.[2]

Other stories concerning psychological stress and their lasting consequences abound. One resident and fire victim was told by her insurance company that it would pay her only one-fifth of the total assessed value of her house. According to her friends and aids she took this news extremely hard. Two weeks later her body was found; she had committed suicide amid a crushing bout of anxiety, fear, and depression.[3] As was the case for so many, the fire's burden carried well beyond the fire event itself as the initial trauma beget further anxieties and hardships. Understanding the unique and sometimes fragile nature of individual bodies and psyches can help us disaggregate and honor diverse and interlinked forms of vulnerability, sensitivity, and coping capacity in fire disaster areas.

As this story highlights, insurance settlements were a major form of anxiety afflicting many in the fire area. Another resident who lost his home near the fire perimeter described the psychological distress of dealing with insurance companies.

> It's easy [for us] to get out of here, it was easy that day ... so I don't worry about that. Risk has to do with the psychological and emotional maze that you go through on the way to a settlement with an insurance company. So risk has to do with what kind of insurance you have. Do you have to go through a drawn-out settlement process, which will probably happen because consciously or unconsciously, the insurance companies have as a tactic to draw out settlements where a settlement is contested.

The family continued by noting the perverse way that certain social groups were exploited during insurance negotiations, thus making the psychological stress of dealing with an already difficult crisis even worse for arguably the most marginalized and vulnerable community members. "Demographics count. If you're a single woman, if you're a person of color, they'll treat you differently. And we were low income. So they accused us of fraud. How could we live here? Even though we had all the proof in the world."

For others the psychological impacts were much more subtle yet still clearly profound and troubling even months and years after the fire. For one male resident who was in high school and home alone at the time of the fire, the effects of the fire seemed minimal. Or so his family thought. One afternoon during the summer after the fire, the boy's family was watching a Fourth of July parade. Walking through the town center, they heard a number of sirens passing from the city fire department taking part in the parade. A minute later they looked around and were unable to find their son. He had disappeared from view. For a moment they could not figure out where he was. According to his parents,

> he had crossed the road, was in the park with his head down like this [*the mother crunches forward with head in her lap, hands and arms shielding her head*].... We didn't even realize how emotional he was ... [the sirens] just brought everything back.

While psychological impacts from the fire lasted well beyond the event, so too did the physical process of relocating between residences. Physical displacement—in some cases for years on end—was a frequent source of stress for fire victims. This constant moving presented immediate challenges for residents, such as changing neighborhoods (sometimes several times), destabilizing a personal and familial sense of place and home, and altering friendships, work commutes, and school attendance among other life activities. These immediate impacts had longer-term consequences as well. For example reduced professional achievements and compromised educational performance negatively affected work and school opportunities in the years and decades ahead. One resident spoke of the unsettling condition of being a child who "bounced from friends' homes to friends' homes for the first couple of months . . . before we rented places." A high school student at the time, it "was a little tough . . . to be settled [enough] to focus on high school." He continued describing the next several years:

> Initially we just, we lived, like just crashed in other people's houses . . . and we sort of went between two different families' homes in the area. . . . I'd been going to Oakland Tech, and so I went to Berkeley High for my senior year of high school, which was kind of a strange thing to do, because it's like being a freshman in your senior year, which . . . made for . . . a very disjointed [experience].

These and other lingering and in some cases life-altering emotional stresses and traumas impacted many in the fire area. So too did many physical ailments. The most obvious corporeal impacts are cases where individuals lost their lives to the firestorm. Also tragic are the many residents who escaped the fire but spent the following years (and in some cases entire lifetimes) coping with severe burns, bodily disfigurements, and the effects of significant smoke inhalation. In most cases these fatalities and injuries did not reflect individual negligence or poor decision making. Rather the area's narrow roads, inadequate water supply, dangerous power lines, and fast-spreading fire front presented conditions that rendered even the most well-prepared resident highly vulnerable to the wildfire. Some of the most persistent fire-induced illnesses occurred after the fire and affected both those who lost their homes and those who did not. One female resident described life in the fire zone in the years after the wildfire.

> Every time it would rain it would activate all of the carbon and all the other stuff that had melted and burned; [it] was reemitted back into the air. There were a lot of toxins that people have in their homes. . . . You have paint supplies, you have kinds of toxins that you have in your garage and in your car

that just got into the air, and we just sat here, lived here, and breathed all that stuff. I got chronic fatigue syndrome about six months later, and was sick for years, and went on disability from work.... I think it definitely affected me, and a lot of people.

Life after the fire was difficult for most everyone. Residents whose homes were spared by the fire's capricious movement and others who lived along the fire perimeter also described living with what some have labeled "survivor's guilt." A homeowner who managed to sneak into the burn area after the fire described his experience grappling with home survival amid a surrounding sea of devastation. This resident explained his attempt to put out small hot spots within the community rubble as other neighbors ventured into the area to check on their homes.

> There were no fire trucks here, because this was designated a perimeter, a no-regress perimeter, while the fire was raging.... But the really sad part was that the families would sneak through, or maybe they got an escort with a car and they would come through, and they would see their houses, lack of houses. It was hard 'cause I was the only person on the street. Then [a neighbor] came by and I was there watering [what remained of] his house down, and I felt like a real shmuck. Here I am, watering his house, so that other houses can't burn. Didn't feel good. Felt selfish.

Others, including the aforementioned woman with chronic fatigue syndrome, spoke about the challenge of voicing her plight after the fire. Because her home did not burn down, she felt it was somewhat insensitive to speak about her hardships even though she knew deep down they were worth speaking about publicly. (Many other residents shared this struggle over how to voice individual trauma without appearing to dismiss or trivialize the vulnerabilities of others.) This dynamic generated an extra sense of stress that only further exacerbated (and in some ways delayed action to treat) her acute health problems. Here again we see the interconnection between different fire-related impacts—in this case physical harms associated with living in the fire area and the emotional stress that arose when positioning oneself, in a comparative sense, as a fire victim. The resident suffering from chronic fatigue described this concern as well as her prolonged exposure in the aftermath of the fire.

You know, a lot of us defer to people that lost everything. I mean, people were writing poetry and telling stories, and it just felt like for me with survivor guilt I felt like, "Well, who am I to complain about anything? I still have my house, I still have my family pictures." It didn't feel appropriate to be telling our story too much, to me. . . . [But] you can't get away from [the illness] . . . it's not like you can sell your house.

Adding to the overall burden of the fire was the significant loss of material items, including personal mementos and irreplaceable keepsakes. Some people lost physical resources and work-related items required to fulfill career responsibilities. This latter group of professionals, particularly tradespeople, mentioned the challenges they experienced working (and earning a salary) in the weeks and months after the fire. A self-employed house painter in the area lost nearly all of his painting equipment. According to his son, his father and mother were "forced to go out and spend a lot of money that initially they didn't actually have. Just to get back some of the basic equipment he had to have to keep working." Drawn-out negotiations with insurance companies delayed property replacement efforts so that many homes wound up paying out of pocket just to put money back *in* their pockets. This financial insecurity compounded an already stressful situation for many working-class members of the community who were living paycheck to paycheck.

But most often it was not high-value or professional objects that generated the greatest sense of loss—insurance companies were eventually able to replace those items. Of much greater concern was the permanent loss of items with considerable symbolic and nostalgic value, things created by and for residents that were deeply personal and irreplaceable. The most profound losses came with the destruction of items that contributed to a fire victims' sense of purpose and self. Photographs certainly fall into this category, particularly because the fire predated digital imagery and remote storage techniques. For household after household entire family albums were lost along with the memories they evoked. A male resident who was a teenager at the time of the fire remembered how his burgeoning interest in photography was impacted by the loss of his home. He described how this meaningful and cathartic pastime consumed much of his youth and contributed to his adolescent identity and maturation. It was his passion and personal enterprise. He recalled that learning how to become a photographer was something he aspired to do more seriously as he grew older. Several of his images were framed around the house and captured his evolution as an artist. He described the loss of these photographs to the fire.

> It sounds a little contradictory, but physical material things bought and purchased in the end didn't really matter ... it's those mementos and self-created things that I think is really what you miss, not you know, just something purchased ... I was baseball obsessed at the time and really into photography, at spring training and taking photographs of a lot of my favorite A's players, and had some of them even sign the photographs—that's one of the few things I still think about ... I wish I'd gotten those out. You know, it's not all these other things that are much more valuable; [the photographs were] just something I created.

The fire area and its residents are filled with countless stories of irreplaceable loss: senior citizens who lost half-century-old, handwritten love letters from their spouse; a couple who had a large portion of their original Native American art collection (several decades in the making) destroyed by the fire; and a female college student who lost over twelve years of daily diary entries—volume after volume of deeply personal writings accompanied by pictures and detailed audio recordings. In these and other cases the destruction of inimitable "things" meant parting with a piece of themselves and their personal history.

For another area resident who lost her home, the fire symbolized the loss of more than just material items; it meant relinquishing efforts to honor the death of her daughter several years earlier. Prior to the fire this middle-aged woman had lost her beloved child, who was in her early twenties at the time, to health complications associated with anorexia. To celebrate her daughter's memory, pay tribute to her passing, and help other parents grapple with children's eating disorders, the mother decided to write a book discussing early illness warning signs. The complex and very personal project was in full swing at the time of the fire. Roughly 75 percent of the book was written and countless hours of audio recordings with her daughter were captured on tape. On the day of the Tunnel Fire, as flames quickly spread through the neighborhood, her family was unable to retrieve these items from the house. There was no digital version or second copy. In a matter of minutes years of work and hundreds of memories and insights were destroyed. This monumental work to honor her daughter's life and help the family members of other victims was gone forever. The effort to productively engage with tragedy was itself a victim of tragedy. Reinitiating the project proved too painful and logistically difficult given the near total loss of primary data. The book was never written.

LIFTING ALL COMMUNITY BOATS: COLLECTIVIZED VULNERABILITY REDUCTION

A heterogeneous and interconnected mosaic of risks and visceral vulnerabilities abound within the fire zone. While the area may hold an overall image of privilege and a general aura of affluence, the hillside community also contains diverse expressions of vulnerability—manifestations that follow levels of hazard exposure, psychological and physical sensitivities, and financial response capacities. And yet these same residents clearly gained from their aggregate, community-wide access to financial and political capital. When viewed as a collective, the fire community was able to combine its resources to secure neighborhood improvements and effectively reduce future fire risks in a manner that might prove more difficult in less affluent, historically marginalized areas. Illuminating household-scale experiences with the postdisaster recovery process highlights particular individuals, community dynamics, and political linkages that actually helped safeguard this area from acute forms of short- and long-term risk. In short, while fine-grained analysis of households can productively illuminate diverse expressions of vulnerability, it also reveals community attributes and behaviors that reduce other modalities of risk.

In the case of the Tunnel Fire the postdisaster response was heavily shaped by the fact that over three thousand homes burned at once. As one resident put it, "If your house is going to burn, be sure that it does with three thousand of your neighbors' in a major media market." In general, a "community voice" tends to get louder and more influential as the population of involved homeowners grows. If the squeaky wheel gets greased, then the loudest and most squeaky wheel gets greased first and most extensively. The sheer number of individuals with a shared and vested interest in the reconstruction process certainly contributed to the relative power of this community response. Of course the total population size of an impacted community only goes so far when influencing the magnitude and direction of public and private investment. Arguably more influential are specific forms of social and political capital—crucial levers of power—that enable the procurement of such investments. This was certainly the case in the Tunnel Fire area where a handful of engaged citizens generated a ready supply of city planning connections that, once leveraged, wound up benefiting all members of the community.

Public participation in the United Policyholders (UP) program illuminates this integrated process of *collectivized vulnerability reduction*. As members of the fire community—faced with a myriad of personal, legal, and financial decisions—struggled to regain footing in the weeks and months after the fire, many residents worked with UP, which was at the time a fledgling insurance holder advocacy program. United Policyholders helped uninformed residents engage with complicated and often adversarial insurance settlements. Not only did UP provide advice to households during insurance company negotiations; it also worked alongside residents to generate data on socioeconomic settlement trends. A community member who worked with UP described its collaborative research, which revealed among other things "that single women did worse than married women, and minority single women ... did the worse of all." As a result of these efforts many residents credit UP with helping them navigate the complicated settlement process and emerge from the ordeal with a just and fair outcome.

The relationship between UP and area residents was mutually beneficial. For United Policyholders the close involvement of residents—many of whom spent months collecting and analyzing area data—was crucial to its success and ascendance into a leading national insurance holder advocacy organization. Indeed UP notes on its website that the "spark for UP was an urban area wildfire that destroyed 3,000 homes in Northern California."[4] While the staff at UP were viewed as crucial allies by community residents, UP itself owes much of its success to the gritty determination of a small group of tenacious community members.

In typical collectivized risk-reduction fashion the diligent commitments of these residents fed right back into the community at large. As

United Policyholders grew in size, publicity, knowledge, and financial capacity, it was able to deliver valuable assistance to more and more residents in the Tunnel Fire area. By helping themselves, active residents helped generate data, insights, and settlement precedents that would eventually benefit hundreds of others in the area. The actions of a few motivated and vocal residents, working alongside (indeed spearheaded by) a small group of committed UP professionals wound up increasing the availability of recovery benefits to the entire community.

Channeling Power: The Political Economy of Power Line Reconstruction

Throughout the postdisaster recover process numerous examples can be found that illustrate how general neighborhood improvements were secured through collectivized vulnerability reduction. These include efforts to attract city investments for the purpose of improving neighborhood lighting systems (many of which were upgraded to resemble graceful, old-world streetlamps) and for modernizing water infrastructure among other municipal upgrades. Here the active participation of a few individuals raised all neighborhood estate values and reduced future levels of fire risk for both residents and firefighters alike.

The case of fire-zone power lines illustrates this process. Aboveground lines presented a serious challenge for both residents and fire response efforts. As later chapters will illustrate in greater detail, exposure to electrical currents resulted in injuries, fatalities, blocked evacuation routes, and impeded firefighting capabilities. Moreover as power lines were destroyed, many water pumps in the hill area failed. As a consequence tanks sitting upslope from city water lines quickly ran out of water once pumps stopped working. In the fire's aftermath most of the power line network was either extensively damaged or completely destroyed. Charred cables and other electrical equipment lay strewn across roadways and draped over cars, trees, and remnant home foundations. Replacing this utility infrastructure would prove to be neither cheap nor straightforward.

Upon receiving replacement value for electricity lines, power companies initially planned to install poles, transformers, and lines similar to prefire conditions. According to one community member active in the negotiation process, neighborhood members collectively replied, "No you won't!" This resident, who was active in the negotiation process, provided a description of resident arguments.

> You cannot just replace what burned. You've got to do it better. Because [prefire conditions] created all kinds of problems for us. Among other things, PCBs (printed circuit boards) . . . I had PCB [material] laying there along those transformers, laying right down on the corner of the street. . . . And this is very unsafe. And not only do power poles burn down. . . . They were exposing chemicals that were very toxic. So there's no way you're going to replace the current infrastructure.

Residents were undeterred when told by city officials that money was simply not available for proposed upgrades. The same active homeowner paraphrased the interaction and the vocal group's response: "Well, look it, I don't tell you how much money you don't have, but if you don't do something about it, a lot of people are going to hear about it. You just can't replace what you got." He continued:

> [The city] knew that that was the right thing to do, so they were having their own meeting, which I was not attending. [They were] saying, "Look, these guys are pissed, we need to do it better." So then we would talk amongst ourselves and say, "It's really important that we get this underground. I'll give you reports or whatever you want and . . . a lot of leaders . . . were saying the same thing." And so we decided that we were going to share on a tax levy on our property, a third, a third, and a third: we would pay a third, the city would pay a third, and [Pacific Gas & Electric] would pay a third.

Although the precise payment mechanisms and allocations shifted over time, this negotiation illuminates the influential role of affluent community activists. Various social advantages, including access to "a lot of leaders" and the willingness to pay a third of undergrounding costs, resulted in the consolidation of significant power and control over planning decisions in hillside community groups. Throughout this multiyear negotiation the financial incentives and benefits derived from these infrastructure upgrades would provide ample motivation for homeowners. As one chief community negotiator noted, power line improvements "would improve the value of our homes. Let's get real here, it was good for us."

In the postdisaster landscape, "power lines" served as conduits of both electrical and political power. They reflect both the physical infrastructure that connects homes, generators, pumps, and substations, and the lines of political influence that connect affluent communities to key decision-makers around the city—decisions that ultimately resulted in the placement of electrical lines underneath neighborhood roadways. In both a material and political sense, then, postdisaster "power lines" have reduced future risk levels, increased property values, and further concentrated affluence and privilege in the Tunnel Fire area.

"Fight Like Hell" and the Importance of Taking Matters into Your Own Hands

To be sure, while collectivized action was certainly important, it is also important to note the tremendous influence of households simply acting on their own behalf. Individual and sometimes *individualistic* actions still mattered in the recovery process. Dogged homeowners with the will and fortitude to stand and fight for what they believed was rightfully theirs should not be underestimated or somehow minimized as an influential part of the recovery process. Underlying the groundswell of collectivized risk reduction and sway of particularly well-connected community members were resolute individuals and families who charted their own postdisaster recovery pathways and secured personal estate benefits on their own terms.

One such couple described themselves as the kind of family that insurance companies thought were "an easy pick" based on a "perception of class," and the kind of people "who wouldn't fight like hell." Speaking about the settlement and reconstruction process, the female partner described her and her husband's determination to fight for their family.

> You don't fuck with me. I have rights, I belong here whether you think so or not. And so we fought like hell, and we didn't give up. And the same thing [as with the insurance companies], we went through hell with the contractor . . . there was this old boys' rule of how we all give in. And I said, "No, you fucked this up, this is what it costs, you're going to pay. End of discussion." . . . But I was . . . the "closer" . . . [and] we got what we were claiming. [Just] 'cause I didn't grow up middle class. I didn't have money, but I'm still an educated person who has a sense of her own value and place in society.

This account like many others from the fire recovery zone indicates the importance of household level resolve during the reconstruction process. Many others in the area who did not have the desire or means to fight with such intensity were ultimately left without some of the personal and estate resources they rightfully deserved.

As these accounts from Tunnel Fire residents indicate, the relationship between vulnerability and affluence is extremely complex. On the one hand, vulnerabilities are variegated at the household scale, as each resident encountered the fire and its effects differently. A nuanced and fine-grained understanding of vulnerability challenges the notion that merely being part of a "wealthy" neighborhood will automatically narrow the scope of potential loss and trauma. The stories of many residents are simply heartbreaking and should never be dismissed or trivialized. Yet on the other hand, the process of collectivized vulnerability reduction does in fact enable entire neighborhoods to benefit from the actions of a few well-connected individuals. These interlocutors of good fortune (alongside the willingness of homeowners to fight on their own) should also not be minimized. Collective action helped the community (and by default each area resident) obtain substantial estate-based benefits during postfire reconstruction. But while it is clear that overall neighborhood privilege aided the recovery process, these improvements will never be able to undo the profound losses experienced by many.

PART III How the West Was Spun

DEPOLITICIZING THE ROOT CAUSES
OF WILDFIRE HAZARDS

6 Smoke Screen

WHEN EXPLAINING WILDFIRES
CONCEALS THE INCENDIARY

High fire risk communities in the East Bay hills are the by-product of a long and complex history of land use planning, resource management, and suburban development. Early on, during the late 1800s and early 1900s, the Tunnel Fire area was transformed—through an interconnected set of activities tied to lucrative resource extraction and in subsequent decades realty speculation, afforestation, and home construction. Many logging paths in Oakland became arterial roads populated by municipal infrastructure, flammable tree cover, and eventually a vast collection of new residential communities in already high-risk fire areas. These are some of the historical processes that constitute the affluence-vulnerability interface (AVI) and highlight the Incendiary at work. Also playing a crucial role in the transformation of the East Bay hills were patterns of government retrenchment, conservative homeowner politics, and state tax restructuring spanning the 1950s–1980s. In the face of a massive reorientation of investment and wealth into suburban landscapes and in order to generate new sources of tax revenue, city officials in Oakland pursued large housing developments within this active fire zone. These are again processes that are comprised in the AVI and that elucidate the Incendiary.

The story of the East Bay hills is a story that repeats itself again and again across the American West: cities become metroplexes expanding outward as subdivisions leapfrog into formerly agricultural, open space and undeveloped areas. And while each story is certainly unique, two things remain constant. First, prospective suburban landscapes are lucrative landscapes and their conversion into residential areas generates both wealth and risk for residents as well as for stakeholders near and far. Second, this is a trend—propelled by considerable financial incentives— that is most often *deflected, concealed, and forgotten in popular and scientific accounts of the "flammable West."* It is this second point—the formation of a smoke screen in front of the core social drivers of fire risk—that the following two chapters seek to highlight in greater detail.

This process of obfuscation is complicated, and no simple answer exists to fully explain the mechanisms of neglect. And yet a close reading of several mainstream debates, conventional explanations, and taken-for-granted truths reveals the ways truncated knowledge is developed, accepted, and propagated within political and scientific discourses on western fire. The provocative article introduced in Chapter 2—"The Flammable West"—like so many others, is a good example of this normative reporting, depicting the West as a victim of an unfortunate set of environmental contingencies. In these discussions damaging wildfires are often neatly packaged and explained to interested parties as being the result of a "perfect storm" of unruly circumstances and unlucky events. The public's understanding of fire and fire risk occurs through a complicated and profoundly tacit process of issue depoliticization—a process that diminishes the role of economic growth and material accumulation that contribute to the formation of the Incendiary.[1]

FROM FIRE TO FIRESTORM: ON THE SOCIAL ORIGINS OF A FIRE CATEGORY

The Tunnel Fire is commonly referred to as the Oakland Hills Firestorm because of its immense size, heat intensity, and high winds.[2] But this label raises an interesting question about the ontology of "firestorms." What exactly are they? Upon investigating the term's origins, one thing becomes immediately clear: there is no such thing as a *natural* firestorm. Quite the contrary, "fires" are only "firestorms" when society says they are. Firestorms are social constructs that we have for many decades now defined, classified, suppressed, created, feared, and managed. If it can be said that humans mastered fire, then modern humans have surely mastered the invention of firestorms.

Rather than focus on how firestorms exist in the world, it will be helpful for a moment to examine how they exist in the world of our ideas. The vernacular shift toward the use of "firestorm" has arisen over time, although the precise origins of the term remain difficult to pin down. Etymologically many historical records show that the term was used frequently during World War II to describe the conflagrant outcomes of massive firebombing campaigns in Japan, Germany, and other parts of Europe. In fact, as discussed in Chapter 5, it was not uncommon after the Tunnel Fire to hear language comparing the postdisaster Oakland Hills landscape to that of postdisaster Hiroshima. Although many historical fires (such as after the 1906 San Francisco earthquake) have generated conditions that would easily satisfy the socially constructed physical properties of a firestorm, it appears the term itself was not applied in any mainstream capacity until the early to mid-twentieth century.

The origins of the term *firestorm* within military contexts are not without consequence. This legacy can be traced through to contemporary fire terminology that connotes the catalyzing source of fire as exogenous, threatening, and uncontrollable. Today the menace of falling bombs onto target landscapes and the resulting "firestorms" they create is being replaced rhetorically by the imposed, out-of-our-control threat of climate change, increasing aridity, and lack of falling rain throughout large portions of the West. Causes of wildland-urban interface (WUI) fires that are internal to impacted environments, such as the presence of extensive home developments, are rendered as victims of these larger external threats. As we will see below, these rhetorical framings portend a more general problem: the concealment of the social causes of damaging firestorms—namely the placement of homes in susceptible landscapes and the profitable industry standing behind those actions. This framing results in the depoliticization of factors influencing pernicious urban sprawl and costly fires (i.e., the Incendiary), and the *naturalization* of wildfires (and firestorms) as simply an unfortunate by-product of the inviolable threat of global climate change (or rogue incidents like unattended campfires and discarded cigarettes).

Increased Fire Risk, Science Classifications, and the Surreptitious Naturalization of Firestorms

A firestorm is defined as "a fire which creates its own weather."[3] This occurs "when the heat, gases, and motion of a fire build up" pulling "air into the base of the fire," leading to towering convection columns that "result in long-distance spotting and tornado-like vortices."[4] For a firestorm to be generated, sufficient fuel load is required that will ignite several adjacent fires in a large area. When these multiple sites of ignition coalesce they become a single firestorm, generating sufficient updraft to create swirling winds and the formation of large pyrocumulous clouds overhead. Unlike forest fires, firestorm designations are typically applied to areas containing a mix of fuel types that include human structures. The high speed, shifting winds, and potential for lightning outbreaks can function as positive feedbacks propelling the growth and damage of a firestorm.

This firestorm definition and its widespread use as a conceptual construct, scientific category, and distinct and observable "thing" has occurred because social risk thresholds are constantly exceeded in various socionatural landscapes, what Ulrich Beck refers to as a "risk society," where planners and policy-makers generally accept the production of new and emergent risks as part of the price of participating in modern-day capitalism.[5] And although many acknowledge and even assume these risks and vulnerabilities, most societies and governments still remain largely fearful of them. Here whole economies and scientific communities are erected for the purpose of studying and managing risks such as those associated with large and catastrophic fires. (Indeed part of the reason why such risks are increasingly accepted in contemporary society is precisely because of the lucrative opportunities that surround their management!)

The rhetorical shift from fire to firestorm thus emerged within a particular social context where fires were (and continue to be) deemed "out of control" and a threat to nearby social assets. And although "out of control" fires could be viewed as perfectly normal in other historical contexts, fire scientists and management officials continue to elevate the significance of the condition-formerly-known-as-fire in response to society's growing anxiety over them. They threaten our viewsheds and the aesthetic appeal of our natural surroundings. They get too close to us. They burn our property. And they threaten our lives and livelihoods. They are not simply fires; they are menacing firestorms. They are fire *disasters*.

Before jumping ahead, let's be clear: these are not *natural* disasters. There is nothing natural about firestorms. We define them. We fear them. We often create them. And we certainly make them more costly. They are man-made disasters. The only thing nature provides is the bright lighting, color accents, and some of the pyrotechnics. Remove the people, structures, and property from the landscape and you have a fire. Put them back in and you have a firestorm and disaster-in-the-offing.

For a fire to be a firestorm it must be sufficiently large and intense. This is a crucial point because it illustrates how we map onto our firestorm designations particular measurable attributes such as fire size, fuel type and diversity, wind speed and direction, pace of spread, vertical development, and the like. However this classification process does something else, something rather more powerful than produce a neat delineation and classification of fire. Presenting the term *firestorm* as a scientifically legitimate category has a surreptitious effect by cloaking it with a sense of authenticity, as if it were something real, natural, and inexorable. But in truth such efforts to classify "fires" simply reflect society's increased proximity to them, sense of threat from them, and need to order and retain control over them. This is the contingent ontology of a firestorm: it does not exist without our classification of it.

Duraflame Firelogs: Constructing the Environment We Fear

Perhaps more crucially, "firestorms" and "disasters" are conditions we also produce in a material sense. We now have a better understanding that the conceptual origins of each are socially constructed (i.e., an area burning at x rate of vertical updraft is a fire, but y rate of updraft is a firestorm; or the same fire may be a disaster to one person but given available resources not to another). However even these rhetorical deployments are dependent on some actually existing biophysical conditions—material attributes that are required to substantiate our observation claims, measurements, and firestorm designations. And these material conditions themselves are shaped by human behavior, community actions, government decisions, and market transactions.

The physical environments—in which sparks ignite and flames spread—are forever altered by the presence of humans. This is characteristic of the Anthropocene, our current—according to many scientists—epoch connoting the unquestionable role of humanity as the main driving force behind the transformation of earth's ecosystems and underlying physical processes.[6] In many exurban landscapes for example humans have entirely altered the physical landscape, often by removing vegetation, replacing it with other land cover, introducing various urban infrastructure such as roads and power lines, while lining the whole region with high-density combustion boxes—customarily referred to as houses. Through this transformation we effectively add fuel to the fire, thereby producing the very kind of fire(storm) we are so fearful of.

It is worth pausing for a moment to elaborate on the notion of homes as combustion boxes. In many ways homes, when placed in areas susceptible to fire, are like Duraflame logs or other specially designed compressed and extruded fire logs. Such logs are primarily constructed out of wood (sawdust) and hydrocarbons (petroleum-based paraffin wax). Most homes are composed of a similar, albeit more varied material composite. Nearly all contain some combination of wood sidings, roofing materials, flooring, interior beams, and other wood-based support and framing infrastructure, while also containing a significant number of petroleum-laced materials such as furniture, carpets, paints, staining materials, and water sealants. Two other similarities between homes and fire logs are noteworthy, if not slightly unsettling: they are both extremely combustible once ignited and they both assist the growth, spread, and duration of a fire (see Figure 14).

The Oakland Hills area, like so many areas across the West, contains a long history of fires and yet has witnessed rapid population growth comprising extensive residential developments. This active historical fire regime does not mean the area was constantly riddled by historical fire *disasters*. Quite the contrary, natural fire cycles helped restore native

Figure 14. Manufacturing and naturalizing the flammable conditions we fear. In many ways homes are a lot like compressed/extruded fire logs. They are heavily composed of petroleum and wood products, are highly combustible once ignited, and assist fire growth and spread. Disasters like the Tunnel Fire arise and are manufactured through the construction and placement of flammable, Duraflame-like objects on the landscape. And yet we naturalize contrived, residential fires by using scientific categories such as "firestorm," which diminish the role that humans play in creating these risks. The unnatural condition of injurious and costly fire events is rendered simply a part of the natural order of things.

vegetation and maintain long-term ecological viability within the region.[7] The recurring presence of fire should not be blamed for the inconvenient and damaging fire patterns that threaten modern-day residential communities around the West. There is nothing disastrous about fire itself. For areas like the Oakland Hills we up the cost of fire. We insert private properties and construct flammable assets. We impose market values and increase exposure. We create fire victims and cultivate loss. In short, we manufacture the disaster.

The undesirable implications of fire increase when human populations and all our trappings are placed within the eventual (yet not unanticipated) spatial extent of fires. We exacerbate fires and oftentimes increase their range and intensity by introducing more combustible material (homes and other structures) on the landscape. We then naturalize these events by developing labels and empirically supported (i.e., scientifically credible) categories such as "firestorm." This scientific and mainstream labeling has the effect of diminishing the very political role humans have in creating these events and crises. The decidedly *unnatural* condition of damaging and costly fire events appears to simply be a part of the *natural* order of things, when in fact there exist many financial incentives and social demands that facilitate their formation. The systematic production of economic benefits generated from attempts to mitigate these risky landscapes is thus able to proceed as simply a logical response to these seemingly inevitable disasters.

THE SHIFTING DEBATE OVER PINE BEETLE AND WILDFIRE ACTIVITY IN THE WEST

As the effects of climate change take hold and thrust the American West into an altered set of socioecological relationships, or "new normals," the edifice of science-based fire management has tried to keep up. The case of the mountain pine beetle (*Dendroctonus ponderosae*) exemplifies the complex, dynamic, and misunderstood nature of fire, its causes, and its interactions with land and life in the rapidly changing West. The devastation of western forests by the mountain pine beetle has been one of the most significant recent ecological changes in the western United States.

Over the past fifteen years mountain pine beetles have caused dramatic tree mortality stretching from Alaska to the American Southwest. Within this massive swath of land, beetles have killed roughly 25,000 square miles of forest, an area nearly as large as Lake Superior. The mountain pine beetle is not new to the West, however, and recent infestations should not be understood as part of a regional-scale invasive species assault. In fact historically the pine beetle has played a crucial ecological role throughout western landscapes by quickly killing off degraded trees and expediting the growth of replacement forest cover. Recent beetle kill patterns therefore mark a deepening and intensification of an already existing phenomenon.

Pine beetles are small—only 5 mm in length—with hard black exoskeletons. They spend nearly their entire lives burrowed beneath the bark of pine species such as lodgepole, ponderosa, and whitebark. As they tunnel underneath bark to lay eggs, they also facilitate the introduction of blue stain fungus. This fungus prevents host trees from repelling the pine beetle through normal defense mechanisms including the production of tree pitch flow which would normally slow or deter insect infestations. Moreover the fungus blocks water and nutrient transport within infected trees. Once attacked, host pine trees are killed within a matter of weeks by the twin processes of larval feeding and fungal colonization.

According the U.S. Forest Service, the rise in forest destruction at the hands of the pine beetle is a result of two factors. First, warmer winters have failed to adequately control larval beetle populations through deep-freeze winter die-off while simultaneously lengthening the beetle-breeding season. This longer reproductive season has a startling secondary effect: studies show that in certain areas pine beetles have shifted from reproducing once per year (univoltine pattern) to twice (bivoltine pattern), thus dramatically increasing the rate of beetle population growth. Second, decreased precipitation has led beetle populations to rapidly spread from already impacted forests into healthier tree stands—particularly at higher elevations and latitudes. Increased aridity and associated tree decline has thus made more forests in the region susceptible to the effects of beetle infestation.

Conventional scientific explanations—and thus resulting management strategies in national, state, and locally managed public lands—have presumed that beetle-infested landscapes increase the frequency of major fire events around the intermountain West. Vast stretches of forested land comprising dead, dried-out, and flammable wood fuel provide the perfect tinderbox environment for increased wildfire activity. As an incident commander with the Rocky Mountain division of the Beetle Incident Management Organization remarked, "When trees die, they go through a natural process of drying," which has been thought to leave impacted areas susceptible to fire. However, "when beetles get into a forest, they essentially shortcut that first stage," expediting the whole process.[8] Campaigns to aggressively combat beetle infestations have therefore used wildfire mitigation as a major justification for increased funding. The 2014 Farm Bill is a good example, as it allocated $200 million to reduce insect outbreak and subsequent wildfire activity across roughly 70,000 square miles of national forests throughout the West.

Reframing the "Underlying Drivers" of Wildfire and the Politics of Privileging Climate Change Adaptation

A 2015 study in the *Proceedings of the National Academy of Sciences* (*PNAS*) paints a different picture. This study out of the University of Colorado Boulder refutes the assumption that bark beetle presence is correlated to fire activity. The study suggests that forests impacted by the mountain pine beetle epidemic are no more at risk to fire than healthy western forests. At the regional level, alterations to beetle-infested forests are not as important in influencing fire activity as overriding drivers like climate and topography. According to the paper, larger western forest fires are occurring within a regional environment experiencing average temperatures that are approximately 2 degrees Fahrenheit higher than in 1970. Moreover a prolonged western drought has plagued the region since 2002 resulting in impaired and desiccated forest environments. As the *PNAS* article notes, "Fire does not necessarily follow mountain pine beetles. It's well known, however, that fire does follow drought." In short the study suggests that forests in the region are already dry enough to promote wildfire with or without the presence of beetle infestations.[9] Due to these recent findings the authors suggest that "policy discussions should focus on societal adaptation to the effect of the underlying drivers: warmer temperatures and increased drought." This policy recommendation is both valid and important and reflects the significance of this study's findings. But it also reflects two ways that the science of fires is unstable, complex, and deeply political.

First, the American West is a region in flux. Entire landscapes, resource stocks, and communities are experiencing rapid alterations in the face of a never-before-seen mix of drought, record temperatures, and rapid population growth. These changes inevitably lead to questions about the nature of important interconnections and feedbacks within ecological, climatological, and social systems—emergent interconnections and environmental dynamics that are oftentimes poorly understood within scientific and decision-making communities. What are the most important drivers and indicators of increased fire activity? Will contemporary firefighting budgets be sufficient to support this projected growth? These transformations and the questions they raise are in turn generating new scientific explanations to assess the nature of these changes and why they matter. And because our desired ability to explain and manage the changing West hinges on this science, these explanations are always debated and contested. The following truism rings clear: As greater reliance is placed on science to address important policy decisions, the more political that science becomes.[10]

Second, this study reflects a deeper and more engrained epistemic commitment: the tendency to rely on adaptive measures when dealing with changing environments (i.e., efforts to adapt to increased flammability). This discursive approach to environmental change gives primacy to overarching natural system drivers that (despite our best efforts to cut greenhouse gas emissions) persist in a manner that is largely out of our control—at least in the near term. For example a multitude of studies suggest that climatic shifts and corresponding surface weather changes are leading to a lengthening of global wildfire seasons that may impact various social systems.[11] In response to a 2015 *Nature Communications* article making this argument, a coauthor importantly but also predictably mentions that "while it may be challenging to slow the pace of climate change, our work highlights hot spots where land management might be able to focus adaptation efforts."[12] What this somewhat fatalist argument conceals however are the complex social drivers that *also* contribute significantly to many regional forest disasters and increase our exposure to them—processes that we are in fact very capable of directly addressing (i.e., efforts to treat the Incendiary). As the authors of a 2014 Natural Hazards Center report mention, "Discussion prioritizing the physical hazard in the case of climate change overshadows and distracts from exposing and confronting the real causes of harm. These root causes remain the many socioeconomic and political processes that push people into vulnerable situations." The authors go on, "Many continue to consider climate-related disasters as exogenous and extraordinary events to which people have to 'adapt.' They do not understand that disastrous consequences are co-produced by the interplay of the social fabric and the extreme climatic event."[13]

To be clear: The conversation emerging from the pine beetle study is extremely important, and the ability of environmental change science to inform policy in direct and practical ways is valuable. Climatic factors *are* highly significant drivers of fire. But the chronic exclusion of factors maintaining and reinforcing what I have described as the affluence-vulnerability interface is all too common. The *PNAS* article on mountain pine beetles suggests that "the government needs to be smart about how . . . funds are spent based on what the science is telling us." Unfortunately, like many studies on fire, the social drivers of landscape change and fire risk and the vast network of profits that undergird them are very rarely directly challenged by the environmental science community (see Figure 15). If money is meant to follow the science and the environmental science community is far less vocal about *mitigation* strategies designed to challenge lucrative land modifications and the production of increased fire risks and costs (not just fire itself), then climate change and associated adaptation measures will continue to rule the day.

Part of the problem is that in these scientific reports the presence of large home developments does not constitute an "underlying driver" of fires. Rather they are viewed as part of a complex assemblage of social and ecological actors under threat from a monolithic and rapidly transforming regional climate. These human artifacts almost appear to be as natural as the region's massive trees, big sky, and undulating countryside. Of course one can quite easily argue that housing developments are not natural, inert, or passive victims. They are a part of historical development structures and enabling policies that increase fire risks and costs. Such landscapes are not merely flammable; they are part of the Incendiary. To confront the development of homes at the affluence-vulnerability interface is to confront an important "underlying driver" of wildfires and fire risk in the West. Indeed much of our concern over fire activity is a direct byproduct of these highly social landscapes in the first place.

This depoliticization of the Incendiary was evident in an announcement by California governor Jerry Brown in early 2016 to issue a $719 million one-time drought package and an extra $215 million to the state's emergency fund specifically to assist efforts to fight the state's next round of large wildfires. According to a spokesperson for the governor, "The conditions have changed in California to the extent that we should change our budgeting process." Given the many large fires burning across the state in recent years, state officials, according to Cal Fire chief Ken Pimlott, are "starting to realize now that we have to get ahead of this sort of thing."[14] So what are the conditions that have "changed" and that the state needs to "get ahead of"? According to Pimlott these monies are necessary because "some of the most devastating wildfire seasons in recent years" are being directly influenced by persistent drought conditions linked to climate change and its effects across the state.[15] Given this climate-centric description of destructive fires and their causes, money earmarked to fight fire disasters is not surprisingly called the "drought package" around Sacramento and within the popular press.

But this type of policy framing and budget justification obfuscates the other important "condition" that has dramatically changed around the region and that is motivating the state government to set aside millions of dollars to fight fires: the steady encroachment of private developments into formerly undeveloped areas at the urban fringe. Unfortunately this undeniably massive change in the California landscape is all but left out of the public conversation. Although the bill could quite accurately be called the "drought and urban encroachment package," government officials and other special interest groups seem quite content with the current, noncontroversial title.

The 2016 "drought package" further demonstrates the persistent privileging of particular socioecological changes (and the diminishment of others) when explaining and responding to fire activity in the western

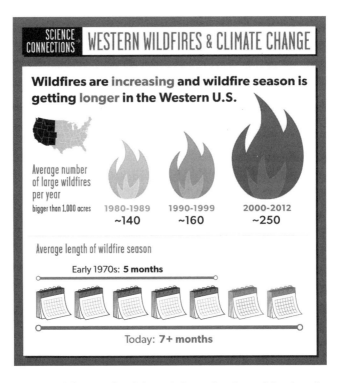

Figure 15 (above and right). An informational panel developed by the Union of Concerned Scientists depicting the relationship between fire and climate change in the West. The panel offers many important and revealing statistics. But this image reveals something else: the minimization of socioeconomic land use drivers and the privileging of climatic forces when explaining the "growing risks of wildfires" in the West. The only reference to residential communities is in the context of adaptation strategies, thus portraying homes as passive victims and not as part of a larger structure of "risk"-producing suburbanization. (Credit: Union of Concerned Scientists, 2013)

Climate change is driving up temperatures and increasing wildfire risk.

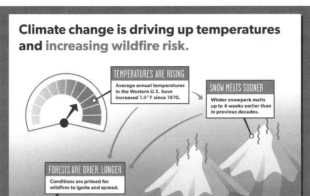

TEMPERATURES ARE RISING
Average annual temperatures in the Western U.S. have increased 1.9° F since 1970.

SNOW MELTS SOONER
Winter snowpack melts up to 4 weeks earlier than in previous decades.

FORESTS ARE DRIER, LONGER
Conditions are primed for wildfires to ignite and spread.

Wildfires are projected to burn more land as temperatures continue to rise.

Projected increase in annual burn area
with an additional 1.8° F rise in temperature

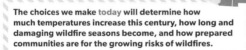

0% — 200% — 400% — 650%

By mid-century, temperatures in the Western U.S. are expected to increase even more (2.5°–6.5° F) due to heat-trapping emissions from human activity.

The choices we make today will determine how much temperatures increase this century, how long and damaging wildfire seasons become, and how prepared communities are for the growing risks of wildfires.

United States. As more science is generated and persistent drought conditions dominate narratives of environmental change, the evolving debate over mountain pine beetles (and other confounding landscape processes) has come to rest in the familiar discursive arena of climate change adaptation. Somewhat predictably efforts by scientists to explain problematic fire activity continue to slip easily from one natural, human-exempted "underlying driver" (pine beetle) to another (increased drought).

THE CONTROVERSY OVER SHAKE ROOFS: THE WOOD SHINGLE AS A POLITICAL OBJECT

A few days after the Tunnel Fire state assemblyman Tom Bates of Berkeley announced plans to introduce legislation banning untreated wood shingles for residential home buildings. The proposed statewide ban would create unified legislation significantly enhancing a patchwork of city and county regulations in place at the time. Mr. Bates knew that implementing such legislation would face an uphill battle as past efforts to regulate wood roofing material proved extremely difficult. Isolated success stories could be found. These included the City and County of San Diego, which established roofing legislation in the early 1980s requiring that new wood shake roofs be made of fire-retardant products. Earlier that decade the California Department of Forestry introduced rules for land under its control requiring new homes to use fire-retardant wood shingles. Several other small cities and municipalities around the state found similar albeit isolated success.[16]

And yet despite these limited achievements strict statewide and city legislation remained an elusive goal. In 1988 Senator Diane Watson of Los Angeles proposed a state ban on untreated wood shingles. The proposal was defeated. She tried again a year later and the proposal was vetoed by then governor George Deukmejian. Nearly every year for a quarter of a century since 1961 former Los Angeles county supervisor Kenneth Hahn pursued a motion calling for a ban on wood shingle roofs. Time and again he was rebuffed by his colleagues amid heavy lobbying from the roofing industry.[17] Despite these persistent efforts much of California—including many high-risk suburban communities—was unable to guarantee the use of proper fire-safe roofing materials in the construction of new housing developments (and the reconstruction of preexisting homes).

Just minutes after Bates announced his intentions to pursue legislation banning the use of untreated wood shingles, he received a call from the Canadian consulate's office in San Francisco. Canada was deeply concerned. As a major producer of red cedar, from which vast quantities of roof shakes and shingles are made, Canada was compelled to voice opposition to the proposal.[18] For the powerful wood shingle manufacturers industry in Canada (as well as in the northwestern United States) this legislation posed a serious threat: another step in a slippery slope toward significantly curtailing or even outright banning the use of wood shingles for home construction. As Bates later commented, "I think they are trying to put out a fire."

Wood shingles and fire have a complicated relationship. Prior to the 1991 Oakland Hills fire it was well understood that wood shingles and shake roofs could increase the susceptibility of homes to wildfire. Roofs and home sidings made of untreated wood in particular were believed to increase both the ignition and spread of fire. For example a 1959 report by the National Fire Protection Association (NFPA) encouraged officials in California and Texas to limit the use of these wood products on home exteriors.[19] Not only are wood shingles prone to easy ignition (compared to fire-resistant composite shingles); they also have a tendency to produce flaming brands that spread downwind, starting new spot fires well ahead of the main fire front. According to fire specialists at UC Berkeley, fires in urban areas with built structures burn much differently than those in forested landscapes. In most cases a moving wall of flames does not engulf homes. Rather embers carried outward from other fire locations get caught inside the nooks and crannies of homes—such as in gutters, under decks, and in unvented crawl spaces. Wood roofs thus hold a double burden: they splinter and generate wind-born embers while also providing crevices for flying brands to land and ignite.[20]

While the roof surface remains a very important factor in determining the rate and likelihood of home ignition, it is now commonly understood that other features such as roof overhangs (eaves) are perhaps just as important if not more. As vegetation and other materials burn along the sides of homes, smoke and embers rise upward and collect underneath the eave. This accumulation of heat and flammable material is trapped beneath the overhang becoming hotter until gutter debris or the building itself finally catches fire. In other words although much attention is given to the top of roofs, homes are just as likely to ignite from their underside if not properly mitigated. It is for this reason that most current building codes now include specific home eave length and ventilation requirements.

Despite the emergence of comprehensive fire-safe home design approaches during the early 1990s, shake roofs in fire-prone areas remained a well-established problem. And yet the majority of efforts to regulate them were rendered ineffective. For decades politicians tried to legislate wood roofing materials out of preexisting neighborhoods (by requesting that all replaced roofs be substituted with fire-safe products) while also preventing them from being used in the construction of new developments across California and the West. For example just months before the Tunnel Fire, in early 1991, then assemblyman Jack O'Connell proposed legislation that would have required a study by the state fire marshal examining possible fire hazards associated with the use of wood shake or shingle roofs. The measure was rejected in its first committee hearing.[21] Even as early as 1923 the Berkeley City Council responded to a major residential fire by passing legislation requiring fire-resistant wood roofing materials. This proposed ordinance was rescinded before it could take effect.[22]

The single biggest source of opposition to this proposed legislation came from the roofing lobby, a powerful triumvirate in Sacramento comprising the home construction building industry, the Cedar Shake and Shingle Bureau, and the Forest Products Association. This was a difficult collective for legislators to defeat. During the late 1980s the wood shingle industry in Canada and the United States employed roughly six thousand people and generated $250 million in wood shingle–related revenue annually. Treated wood products alone generated $50 million in revenue for the industry. With much to lose, members of the roofing industry stood together in collective resistance to proposed statewide bans on untreated wood shingles and more restrictive measures banning wood shingles altogether. Much of the resistance came from the nonprofit organization now identifying as the Cedar Shake and Shingle Bureau (CSSB).[23] Originating in 1915 and rebranded in 1963 as the Red Cedar Shingle and Handsplit

Shake Bureau (then renamed again in 1988 as the CSSB), the organization represents hundreds of manufacturers of roofing products in various trade, business, and legal sectors. Again and again probusiness legislators sided with manufacturers and the CSSB to protect the financial well-being of this massive and highly profitable roofing industry.

The Many Political Dimensions of the Shake Roof Debate

By late 1989 the drawn-out battle over wood shingle roofing in Los Angeles finally neared an end as an ordinance was approved that would prevent the use of all wood shingle materials on new building construction. Years (and in some cases decades) of pressure from legislators like state senator Diane Watson and Los Angeles county supervisor Kenneth Hahn had finally paid off; Los Angeles became the first major city in California to implement a complete ban on wood shingles. Leadership within the Los Angeles Fire Department (LAFD)—who took a very vocal and principled position that wood shake and shingle roofs (whether treated or untreated) are dangerous and more susceptible to ignition than composite varieties—provided a critical voice in support of the ban. The LAFD cited the nearly one thousand homes with shingle roofs that had been damaged or destroyed since the 1961 Bel Air fire as a stark reminder of why this strict ordinance was necessary.

Needless to say the Cedar Shake and Shingle Bureau was in a panic calling the legislation "unwarranted and discriminatory" and "unconstitutional." The president of the CSSB noted, "What's disturbing to us is that everyone's quick to point the finger of blame on wood roofs." After the ruling the CSSB sued the City of Los Angeles arguing that the legislation "unfairly discriminated" against industry manufacturers.[24] To their satisfaction the state attorney general issued an opinion declaring that cities do not have authority to impose fire codes more stringent than those of the state. The wood shingle lobby even tried to promulgate its own science by calling on the Canadian consulate to host a special materials test that would demonstrate the safety of its treated wood materials for city officials. This last-ditch effort went up in smoke. A federal ruling was handed down in 1991 that effectively upheld the ban. While it is true that many cities still lack ordinances preventing wood roof construction, others like the City of Oakland made strategic progress after the Tunnel Fire requiring that all new Oakland Hills houses (and roof conversions) use Class A roofing types made with nonwood, fire-resistant materials including slate, clay, concrete roof tile, or steel shingles.

The case of Los Angeles, like many other cities, illuminates the deeply political and protracted process of modifying the entrenched and economically powerful residential home construction and material manufacturing industries, even despite the presence of unrelenting political pressure to enact mitigative legislation. The battle over wood shingles and cedar shake highlights several important points. First, fire mitigation activities threaten the financial interests of many entrenched interests within the private home construction industry. Wood shingles become political objects that are at once a source of worker livelihood and economic security in the Pacific Northwest and Canada, and also a symbol of reckless and risky home construction activities in fire-prone WUI environments of the American West. For many the debate came to represent a choice: the destruction of cedar shingle homes or the destruction of the cedar shingle industry. (Of course the CSSB would not see the protection of one or the other as mutually exclusive.)

Second, the case of wood shingles illustrates that public disputes over residential fire risk are quite often not about *whether* to build more homes but rather about *how* to build them—and further, *who* gets to participate in the construction. For several decades beginning in the 1960s concerns about reducing vulnerability boiled down to debates over home construction material. The inevitability of the construction itself went largely unquestioned. Thus what's not included in the discussion is what has made this debate deeply political.

Third, moving from discourse to materiality, the case of wood shingles illuminates how powerful financial interests forestalled the passage of risk reduction legislation for many decades and increased fire risk for *actual* residents and first responders in fire-prone landscapes. Thousands of homes were constructed during this period using roofing materials that were known to increase the likelihood of structure ignition from wind-blown embers. Yet despite these known risks profitable home construction sped right along. This process highlights how profits for wood shingle and cedar shake manufacturers and elevated fire vulnerability for urban residents are produced in concert—an example of what I have referred to elsewhere as a process of "vulnerability-in-production."²⁵

All three cases in this chapter—the social origins of the term *firestorm*, the relationship between pine beetles, climate change, and fire, and the drawn-out battle over wood shingle roofs—illustrate the way the Incendiary is naturalized and accepted as simply an inevitable outcome. By establishing a scientifically credible "firestorm" category, socially produced fire disasters appear ordinary, familiar, and strangely natural. Underlying economic drivers such as the role of the home development industry tend to be rendered immutable and thus immaterial. The same holds true when wildfire activity is characterized as a by-product of underlying drivers such as mountain pine beetle and climate change. The political economy of urban growth and suburbanization is left as an unquestioned and unavoidable outcome. And when debates over proper home roofing materials and concerns over residential fire risk begin not by asking *whether* to build more homes but rather by debating *how* to build them, the inevitability of home construction itself goes largely unchallenged. In these ways wildfire risks and hazards in the West—and the influence of the Incendiary on them—are depoliticized.

7 Debates of Distraction

OUR INABILITY TO SEE
THE INCENDIARY FOR THE SPARK

Our difficulty addressing the underlying social causes of increased wildfire risk and costs can be explained in part by a myriad of distracting alternative and proxy debates. There are a vast number of these corollary disputes, which range from disagreements concerning the appropriate role of government in managing private property activities to quarrels over the meaning and intended function of whole landscapes. Despite their diversity these issues hold a similar quality: each functions as a spark that ignites disputes at neighborhood, city, and regional levels. Once communities, managers, scientists, and politicians become mired in these debates, the Incendiary may become less visible, less acknowledged, and seemingly less important. As we labor to put out small fires, we fail to see the whole wildfire complex. We may understand this as a process of issue *repoliticization,* where the emphasis of public conversations on the social causes and implications of fire risk is replaced by other, seemingly more contentious debates of distraction. The consequences of this obfuscation are not insignificant; such disagreements distract us from addressing more fundamental questions that could provide a pathway toward destabilizing the foundations of the very problem we seek to upend.

THE DEBATE OVER EUCALYPTUS IN THE OAKLAND HILLS

On March 5, 2015, the Federal Emergency Management Administration (FEMA) announced its decision to grant $5.67 million to the California Office of Emergency Services to implement major fire mitigation activities in the East Bay hills.[1] Mitigation provisions called for tree and vegetation removal including the thinning of eucalyptus groves—a tree species that a former UC Berkeley professor of forestry and conservation once described as "the worst tree anywhere as far as fire hazard is concerned." The grant application that was approved for funding marks a middle-ground position; it is more exhaustive than basic understory debris removal but less intensive than clear-cutting and complete eucalyptus eradication—the latter finding support within the University of California and some neighborhood groups.

The "unified methodology" endorsed by FEMA integrates the demands of all participating agencies while also requiring coordinated actions that integrate pursuits across jurisdictions. According to FEMA, the project will

> reduce fuel loads and fire intensity, primarily by thinning plant species that are prone to torching, and by promoting conversion to vegetation types with lower fuel loads. In many areas the proposed and connected actions would preserve oak and bay trees and convert dense scrub, eucalyptus forest, and non-native pine forest, to grassland with islands of shrubs.

This decision marks an important development within a long and contentious debate over the management (indeed meaning) of eucalyptus in the Oakland and Berkeley hills area. And yet despite this apparent resolution the decision itself has become a new catalyst for disagreement. On one side the Hills Conservation Network (HCN)—a collection of community members inside the affected area—has discussed suing FEMA because it says too much emphasis is being placed on nonnative tree removal. Rather than engage in clear-cutting (a characterization FEMA strongly denies), HCN president Dan Grassetti suggests "the best thing you could do in this area for fire mitigation is to maintain the tall trees and eliminate the ground fuels and the fire ladder." Meanwhile another community organization, the Claremont Conservation Conservancy (CCC), holds a different interpretation of the plan, remarking that FEMA didn't appear to go far enough in its effort to eradicate eucalyptus. "It's a half-hearted effort," said Jon Kaufman, stewardship coordinator of the CCC.[2]

These positions are a microcosm of a much larger set of debates over the role of eucalyptus and other flammable and nonnative vegetation in the Oakland Hills as well as in other fire-prone regions around the western United States. And while these debates are themselves controversial and divisive, they actually have a more pernicious effect. As currently framed, the debate over eucalypts diverts attention away from broader social processes actually driving fire vulnerability. In so doing, these debates tend to naturalize fire (and our concerns over fire *risk*) in the area as simply "the way things are." Here and elsewhere these problem framings and the debates that ensue (e.g., concerning levels of tree flammability and the merits of nonnative species eradication) wind up concealing the political economy of real estate development that has produced the need for fire-adapted communities and the procurement of multimillion-dollar, taxpayer-funded FEMA grants in the first place.[3]

The debate over eucalyptus and fire mitigation in the East Bay hills is indicative of many land management disputes around the United States and illustrates the aforementioned process of issue depoliticization and repoliticization—where the former involves stripping an issue of a controversial underpinning (in this case processes operating at the AVI) and letting foundational explanations of social-environmental change go unchallenged, while the latter involves the emergence of alter-debates that distract participants from confronting the foundational and structural origins of the Incendiary. In this way repoliticization helps maintain the broader process of depoliticization.

In 1992 the mayor of Berkeley, Loni Hancock, and the mayor of Oakland, Elihu Harris, chaired a report of the Task Force on Emergency Preparedness and Community Restoration. According to the final report, "The high density of flammable structures contributed significantly to the spread and intensity of the Oakland hills fire. Trees did play a role in spreading the fire, but in many cases the trees caught fire from the houses, not vice versa." This was a remarkable statement at the time because it bucked conventional explanations that typically found fault with the presence of nonnative, overgrown vegetation. Yet despite this statement the debate over fire-safe communities in the years and decades to come—including the 2015 FEMA award—focused almost exclusively on the "vice versa." This problem framing is indicative of fire mitigation efforts in general: homes are treated as a given on the landscape and efforts to mitigate fire risk are structured around this a priori condition. This approach conceals the fact that fire (and fire risk) has only emerged as a pressing social concern because of the encroachment of homes into areas that were already susceptible to fire. By omitting the crucial role humans have played in producing these landscapes and instead insinuating, through discursive silence, that homes are a part of the de facto landscape, residential fires simply become part of the natural order of things. While controversies over how to protect homes may arise, the homes themselves are never controversial. And while the Oakland Hills is a well-established area, this same focus on vegetation management here and elsewhere prevents meaningful conversation about the construction of homes in areas that are still undergoing significant land conversion.

This is how the depoliticization of fire mitigation proceeds. The broader political economic causes of fire risk are swept under the debate floor rug—as are the unsavory processes of greed, avarice, and ineptitude that continue to produce development outcomes running contrary to wider risk reduction goals around the western United States. What about the immense profits that were (and continue to be) generated while developing these now high fire risk areas? Not up for discussion. Nor are the many lucrative opportunities that arise as part of contemporary efforts to promote fire-adapted communities.

The tone and content of current debates also distract us from the fact that eucalyptus trees—presented as unique from and even oppositional to nearby home structures—actually accompanied the construction of these residential neighborhoods. As earlier chapters described in more detail, eucalypts were distributed around the eventual fire area in order to increase property values, extract profit, and generate affluence within these landscapes. Thus if one is to talk honestly about eucalyptus, one must speak directly to its residential landscape counterpart, the home. This means shifting the debate so that it more directly confronts the social drivers and attitudes that have placed homes in fire-prone areas—a process that often involves the introduction of flammable vegetation cover. In short, rather than contemplate eucalypts and fire in relation to homes, we should be considering eucalypts and homes in relation to fire.

Unfortunately the current fire mitigation ethos has minimized these concerns and the role of "process" more generally through systematic issue depoliticization. Vegetation cover in the Oakland and Berkeley hills area has provided fertile ground for an (ostensibly) ecological debate that has captured the attention of governments and community activists. (This is exemplary of many other debates including those concerning proper roofing materials described in the previous chapter.) The first part of this debate concerns the relative contribution of eucalyptus trees to overall landscape flammability, while the second part concerns competing perspectives over how to define and conserve native vegetation.

Debating the Contribution of Eucalyptus to Overall Landscape Flammability

For many, eucalyptus trees represent a highly flammable and thus dangerous tree cover. According to the University of California (UC) Berkeley campus environmental manager, who is also a member of the UC Fire Mitigation Committee:

> Eucalyptus are a special risk because they drop tons of dead leaves and branches on the forest floor, litter that provides excessive fuel to fires. Their low branches serve as fuel ladders up to their high crowns, and their volatile oils burn hot and fast. When eucalyptus catch fire, flames shoot up to the crowns, which send embers flying. High winds can carry the embers across firebreaks and clearings and into residential areas.[4]

Moreover the level of tree cover present in the area is considerably higher than a century ago, and before Anglo-Americans settled the area. The images of the Claremont Hotel and Miller homestead in an earlier chapter illustrate this afforestation trend. The obvious goal then should be to remove certain vegetation elements and limit the amount of flammable biomass on the landscape. And eucalyptus, which is both nonnative to the area and infamous for its high resin and levels of ground litter, appears to be a logical place to start. The CCC is quick to add that of all the area that burned in the 1991 fire, the largest percentage of land containing no structures is characterized as eucalyptus groves (21% of total acres). This is compared to 18 percent for northern coastal scrub, 9 percent for Monterey pine, 5 percent for coastal scrub and grassland mosaic, and 3 percent for coast live oak and coastal scrub mosaic.[5]

To others however eucalypts are convenient scapegoats unfairly blamed for the spread of the 1991 fire. They are seen as an "actionable" target for cities and when sufficiently reduced in numbers give the appearance of successful fire mitigation. For community groups like the Hills Conservation Network placing such blame on eucalyptus is both unfair and misguided. These parties question how much of the 21 percent of the burn area characterized as "eucalyptus groves" actually contains eucalypts: is it a contiguous forest belt or merely the presence of one small copse? Furthermore, large disaster events such as the Tunnel Fire—aided by dry conditions, high winds, and plenty of other fuel on the landscape—would almost certainly have spread through the region regardless of how many eucalypts were present. Thus the elimination of these species is viewed as both an overreaction to a fictitious problem and an oversimplification of a much more complex landscape. For the HCN a more measured and advisable approach involves selected tree cutting and the removal of lower branches, understory vegetation, and leaf and bark litter.

Disagreement over eucalypt flammability can even be found in the way combustible biomass content on the landscape is enumerated and reported. For community organizations like the Claremont Conservation Conservancy the massive quantity of leaf, bark, and branch debris lying beneath towering eucalyptus is of major concern. With this extensive accumulation of ground litter, eucalypts appear to place much greater fuel onto the landscape than other tree species. The CCC refutes the suggestion that eucalypts are being scapegoated and cites various reports including a postfire FEMA report declaring that "Eucalyptus and Monterey Pine have been identified as fire hazards and their spread should be controlled." Moreover, according to the East Bay Regional Park District, a mature eucalyptus forest in the East Bay hills contains more than fifty tons of combustible litter per acre, far above fire-safe standards of five tons per acre.[6] Dissenting voices such as the HCN contend that these measurements include portions of the eucalyptus trunk, which are no more flammable than the trunks of other trees and so should be removed from the calculation. Furthermore replacement vegetation from oak and other forest cover combined with chaparral, brush, and sources of ground litter should be figured into these net numbers. Still further, skeptics of the park district calculations suggest that the qualitative nature of forest composition—including species types and horizontal and vertical land cover characteristics—must be sufficiently factored into the analysis to accurately gauge net flammability.

Here we can see how the debate over landscape flammability is in large measure contested around how best to enumerate vegetation fuel load, a process that is itself fraught with inaccuracies, subjectivities, and oversimplifications. And as the community chases scientific clarity on important ecological questions—such as what constitutes a eucalyptus grove or the relative combustibility of tree trunk litter—it circles ever deeper into a "debate of distraction" vortex.

Defining Native

Meanwhile, as the debate over flammability rages on, a second issue has emerged around how best to manage nonnative species in the area. This disagreement has produced differing opinions on what constitutes the appropriate amount of eucalyptus on the landscape. For example, what is an acceptable tradeoff between eucalypt removal and increased native species presence? How much herbicide can be justified in invasive species removal? These questions and others like them all speak to a critical underlying question: What counts as native in the first place? Eucalyptus trees were introduced to the area in the late 1800s, meaning they have been a part of the landscape for nearly as long as western settlers have occupied the area. Put differently they are more native to the East Bay hills than every resident currently living there. No one alive can recall a time without eucalyptus, and one would have to go back at least four generations to predate its presence on the land.

These landscape alterations and subsequent debates remind us that environments are always changing as a result of direct human forces (e.g., land conversion and species introduction), indirect human forces (rapid climate change and species migration), and other nonhuman forces (longer-term climate oscillation). So determinations of what is native and what is not can become a dangerous game of subjective preference and selective historical benchmarking. How far back must we go to determine what is native? And how do we calibrate our answers to this question in light of shifting environmental baselines and an increasing social acceptance of novel landscapes and new ecological normals?[7] Moreover are we more likely to accept an exotic species as "native" if it integrates with, supports, and even enhances local ecological function? The case of the East Bay hills area exemplifies this seemingly intractable debate, one that has been active in California and elsewhere for several decades. In fact environmental activists and governments concerned with the preservation and restoration of native species led efforts to eradicate nearly all blue gum eucalyptus on Angel Island in the San Francisco Bay in 1990. Eucalypts have also been removed from Annadel and Montaña de Oro state parks in Sonoma and San Luis Obispo counties respectively for similar reasons.

Despite the hazy definition over what counts as native, the CCC and others remain steadfast in their pursuit of significant eucalyptus thinning. The CCC's objection to the FEMA plan's unified methodology is largely based on its sense that not enough trees will be removed from the landscape. According to the CCC, "In Claremont Canyon complete removal of these dense tree clusters will enable native species of oak, bay and willow to regenerate." Many politicians, including several Berkeley and Oakland council members, have officially supported native species restoration efforts.[8] To these parties the primary purpose of eucalypt removal is to restore both native plants and fire-safe communities, particularly in the wake of decades-long forest neglect resulting in extensive nonnative afforestation.

The HCN however views the debate quite differently. According to the network, opponents of nonnative species have latched on to fire risk as a way of removing trees like eucalyptus from the area. Fire management is being enrolled rhetorically and mobilized politically to support and assist efforts to achieve a very different land conservation objective—invasive species control. According to the HCN director, in the years following the 1991 Oakland Hills fire "the eucalyptus was scapegoated" despite the complexity of causation that led to the fire's ignition and spread. The director argued, "There's been a systematic effort among several agencies to take advantage of people's fear of eucalyptus to get funding to do what they

Figures 16 and 17. Popular portrayals of eucalyptus. Those in support of only minimal eucalyptus thinning tend to embrace and display images of eucalypts as a cooperative, interdependent, and nurturing hillside tree species (top image). Those in favor of more aggressive tree removal will tend to show images such as large, imposing, and uncooperative eucalyptus groves (bottom image) that seem to crowd out California bay laurel (*Umbellularia californica*) and other coastal hill forest species.

want to do to the land."⁹ In general the HCN is more willing to accept eucalypts as one component of a broader species mosaic on the landscape. As of mid-2015 for example the organization's website home page boasted a large photographic banner of an owl sitting in the nook of a eucalyptus tree (see Figures 16 and 17). As opposed to seeing eucalypts as a threat to the local ecology, the HCN views these trees as contributing to the very environment residents have come to know and love.

The debate over eucalyptus management in the East Bay hills has been largely focused on two controversial issues: tree flammability and native plant restoration. The discursive prominence of these disagreements and controversies, certainly important in their own right, shifts attention away from important cautionary insights that could inform other quickly developing areas of the West. Displaced from the conversation are questions concerning how human populations came to populate the area and more critically the persistent role that profit-seeking real estate development interests have played in creating this fire risk—a process of wealth accumulation that was critically influenced by the introduction of the "problematic" eucalyptus. (These are developments we see playing out across the western United States, only with different landscape features, development patterns, and introduced tree species; see Figure 18.)

The Politics of Belonging: The Battle to Control the Narrative on Eucalyptus and Its Racial Overtones

The debate over eucalyptus revolves around two landscape visions. One perspective positions eucalyptus as a threat to native vegetation cover and neighborhood security. The other sees eucalyptus as just one species in a productive mosaic of vegetation types with a less distinct link to fire risk. This debate however does more than distract us from political economic explanations of landscape change (through issue repoliticization). It also reminds us that debates over landscape form and function often boil down to who can control the narrative of what belongs and does not belong on the landscape. As the environmental historian Peter Alagona reminds us, struggles over individual species "often serve as proxies for much larger debates involving the politics of place," which may be understood as "an ongoing cultural conversation about who should have access to and

Figure 18. Home structures are the fuel load that matter most. During the 2003 Cedar Fire in San Diego an individual home's roof is ablaze while surrounding eucalyptus vegetation is largely unburned. Homes along the rest of this cul-de-sac (out of view) show a similar burn pattern. A recent study in the journal *Nature* (Moritz et al., 2014) suggests that locations with the greatest total area burned correspond with areas containing high population densities rather than areas with high forest cover. In other words, to track elevated fire activity follow the people and homes, not the trees (particularly those set back from home structures). Focusing on vegetation in regions with room for WUI growth can divert attention away from more important topics, such as debates over why new homes are being constructed. (Credit: NY Times and John Gibbens, Union-Tribune)

control over natural resources."[10] In the East Bay hills, disagreement over eucalyptus may therefore be understood as a proxy debate between competing visions, an arena within which minds are shaped and where control is asserted over the meaning and purpose of the landscape for all its inhabitants—humans and nonhumans alike.

Efforts to control the terms of "belonging" in the hill area are certainly not new. If we look back at early development activities, a similar tone and set of rhetorical maneuvers was used to define the characteristics of acceptable inhabitants and to support xenophobic and racist real estate practices in the region. From the outset wealthy, white communities settled these hillside neighborhoods of Oakland. This was a very intentional demographic outcome. As a 1911 Laymance Real Estate Company brochure stated about one neighborhood in the Oakland Hills, "no one of

African or Mongolian descent will ever be allowed to own a lot . . . or even rent any house that may be built there." A very particular type of racial composition (Caucasian) was therefore promoted in the region as African American, Asian American, and Hispanic communities concentrated in other parts of the city. These nonwhite populations were perceived as a threat to this stable and secure residential community created as a safe haven from the increasingly perilous and blighted city below.

For certain groups in the contemporary East Bay hills eucalyptus trees are framed as flammable (read dangerous) and invasive (read do not belong). Nonnatives become the nonwhite. And invaders become the people of "African or Mongolian descent." Of course bringing attention to this similarity is not to suggest that those in favor of native vegetation are racist or bigoted. Rather the point is that the practice of delimiting what belongs and does not belong on the landscape is a very political act. It is a subjective process that requires determining what (or who) is acceptable and what (or who) fits within the "desired" vision of a place. Throughout history this has meant prohibiting the presence of things—plants, people—that threaten this vision. Under current deliberations some support while others challenge the notion that lack of belonging (nonnative) will actually lead to reduced security (flammability). The debate is therefore a contested process of inclusion and exclusion. And it necessarily requires that groups vie for control over definitions of what "belong," "acceptable," and "desirable" should actually mean.

In the post–Tunnel Fire context the debate over eucalyptus may therefore be understood as a difference of opinion over how best to achieve a landscape of productive, trustworthy, cooperative, and interdependent land features. Competing visions of the hillside have led groups to argue over the material attributes and symbolic meaning of eucalyptus trees. Along the way environmental science calculations have been strategically used to justify and substantiate these visions and meanings by reducing otherwise complex socioecological conditions to simple and easy-to-quantify indicators of vegetation flammability, environmental risk, and landscape suitability. Actuating desired management decisions and resulting landscape conditions has ultimately rested on the ability of different groups to control the narrative of eucalyptus using metaphors (or surrogate inferences) such as "dangerous" and "lack of belonging" to justify their removal—a brand of argumentation that is eerily reminiscent of early twentieth-century responses to the perceived threat of non-Caucasians residing in the hill area.

STATE RESPONSIBILITY AREAS: AN IDEOLOGICAL STRUGGLE OVER PROPERTY RIGHTS AND GOVERNMENT ASSISTANCE

In 2011 the Fire Prevention Benefit fee program was enacted within California State Responsibility Areas (SRAs). The program was initiated in order to collect payments of up to $150 for each habitable structure on the property of landowners in areas susceptible to wildfire.[11] A statewide program coordinated by Cal Fire, the fee supports fire prevention activities across California that aim to reduce fire risks for nearby residents and protect state and municipal infrastructure. Program activities include vegetation clearing, firebreak construction, defensible space inspection, hazard severity mapping, and public education activities. State Responsibility Areas are geographically liminal and typically reside outside of official city boundaries. They occupy landscapes where the State of California (Cal Fire) has financial responsibility for fire suppression and prevention and thus do not include lands within incorporated city boundaries or areas under federal ownership. For example in the Oakland Hills area designated SRAs straddle rolling slopes between Alameda and Contra Costa counties primarily occupied by open space and regional parks and preserves. At the program's inception in 2012 the State of California issued its first round of SRA payment requests. Because requests are sent alphabetically, Alameda County's approximately 4,200 eligible residents were first to receive notice (including residents in areas impacted by the Tunnel Fire).

Rollout of the SRA program has not been without controversy. The grounds for disagreement largely mirror those associated with other government initiatives that obtain payments from property owners in exchange for the right to manage some aspect of their private lands. Program proponents argue that these fee-generated revenues are necessary to overcome decades of budget cuts resulting in diminished fire prevention capabilities and increased fuel buildup across the state. For these individuals the SRA represents a reasonable price to pay for the right to live in high fire risk areas. Moreover government-coordinated efforts are viewed as the only way to ensure that consistent fire mitigation goals are achieved across contiguous yet heterogeneous land designations. Advocates argue that the fee follows the "beneficiary pays" principle by asking those directly benefiting from management activities (in this case fire-risk mitigation and improved community preparedness) to take disproportionate financial responsibility for the program.

Dissenting voices question the redistributive nature of the fee and oppose what they see as yet another effort by the government to impose taxes on members of the community. The Howard Jarvis Tax Association (HJTA), which rose to prominence for its support of Proposition 13, for example has raised several issues including questions over what constitutes a habitable structure and what fee collection methods will be used. It also views the SRA as a direct threat to private property rights. When viewed more closely, programmatic critiques all tend to challenge a much broader and more fundamental issue: the perceived unfair redistributive nature of the fee and of government collection programs that are "desperate for revenues."[12] Many see the SRA fee as nothing more than a thinly veiled tax, thus requiring two-thirds vote in the legislature.[13] For opponents, "The [SRA] fire tax is a direct violation of Prop. 13," that is, "taking more money from hardworking people for a program they were already paying for."[14] According to a legislative director of the HJTA, a reapportioning of personal estate earnings for public expenditure purposes is a misleading "general fund shell game."[15]

A Debate Redux: The Wildfire Prevention Assessment District Fee

While these developments unfolded at the state level, another similar debate was brewing in the City of Oakland concerning the Wildfire Prevention Assessment District (WPAD) fee. The WPAD, which is in many important ways similar to the SRA, was first implemented in 2004 to supplement fire prevention, mitigation, and preparedness programs that were historically provided by the Oakland Fire Department. These include services to increase property compliance inspections, public outreach and education, goat grazing, and brush and dead vegetation removal. In 2004 voters within the district boundaries approved the WPAD and associated fee for a ten-year period ending in 2014. In each year a $65 fee was applied to single-family dwellings in designated WPAD areas leading to over $1.8 million in revenue.

Voters in late 2014 were again asked to approve the WPAD fee through an annual parcel tax that would extend the program for another ten years. Owners of undeveloped parcels were asked to pay $39.00, and condominium and multifamily property owners, $58.50. All others, including single-family homes, were asked to make an annual payment of $78.00. To the surprise of many in the fire prevention industry, renewal of the WPAD fee failed to reach the required 67 percent approval rate by just 66 votes. (The 67 percent threshold stands as a result of the WPAD's designation as a tax—a classification that was itself a victory for the antitax community, which understands the difficulty of achieving 67 percent support for most referenda.) Several factors are believed to have contributed to the measure's failure. According to both a battalion chief with Cal Fire and a local city fire marshal, the rejection of the WPAD fee extension can be attributed to a still prevalent antitax sentiment. They noted that many hillside residents participating in the election look unfavorably upon paying "extra" for wildfire mitigation services, and as the fire marshal put it, "People are somewhat fee weary, they're tax weary," particularly in a recession when "their ... income hasn't gone up."

According to these officials, skepticism toward new fees and taxes is compounded by two other factors. The first is a limited knowledge by residents of the history of the place they live in, a condition that inevitably deepens with the introduction of new residents. For these individuals, many of whom located to the area more than twenty years after the fire, the conflagration represents a distant and increasingly hazy event. The same fire marshal noted that many people living in the area "didn't go through the fire. They don't even know what it is all about." Even survivors of the 1991 conflagration mention how wildfire vigilance tends to subside with each passing year as the firestorm gets relegated to "anomalous event" status. For example when asked if he was concerned about the possibility of another large conflagration, one resident responded without hesitation: "We won't get the firestorm. The firestorm is not going to happen again."

A second factor involves the unclear benefits of fire prevention. Unlike easy-to-quantify performance metrics such as the average response time to emergency calls, the total number of people pulled from cars, or the number of homes saved and fires extinguished, the WPAD program is most successful in ways that are difficult to quantify—that is, when nothing happens. Enumerating the successes of ten years of prevention activities is simply hard to do. For example one fire official mentioned:

> When you're on this side of the un-shiny penny of fire prevention, it's very difficult.... When we installed smoke detectors last year, how many people did that save? How many fires did that suppress? Those are numbers that are much more difficult to get.... And so when people start looking at budgets and want to start cutting programs . . . they've got to cut somewhere.

Indeed championing ten years without a major fire is a tough sell, particularly when conflagration-free decades are a frequent occurrence. It would be reasonable for a voter to ask how exactly this is an *improvement* over previous periods. Both of these factors—reduced knowledge and memory of place, and difficult-to-quantify prevention benefits—contributed to a belief among a third of voting participants that the program just wasn't worth the money, particularly as the imposed fee was additional to already existing city tax structures.

The SRA and WPAD initiatives bring into sharper relief a recurring tension between three important developments in the urbanizing West. First, each program highlights how both Cal Fire and the City of Oakland realized that after several decades of stagnant and reduced budgets additional revenue streams would be required to adequately reduce fire risks for nearby communities. In both cases the chosen method for augmenting fire services was through the implementation of designated property-owner fee programs. Second, a pervasive anti-fee sentiment, which can be linked to a wider homeowner antitax movement, has challenged the necessity (and in some cases legality) of these fees. And third, public skepticism and even rejection (in the WPAD case) of supplementary fire service fees presents challenges for agencies seeking to boost fire prevention activities and reduce social vulnerability to fire.

But most importantly this debate signifies another ideological battleground—this time concerning the responsibility of citizens to fund their own protection and the appropriate level of government involvement—that captures the attention of residents because it pulls at a fundamental yet contentious thread of disagreement concerning how to balance state intervention with individual rights. This is a debate that plays out all across the West as governments attempt to levy fire suppression fees onto local residents in order to reduce mitigation costs for city, county, and state agencies. And yet all the while this debate of distraction bogs down other important discussions concerning whether to stop building more homes and stop feeding the Incendiary (and accepting its unavoidable march) in the first place. The compromise position is typically one that allows at least some new home construction (and consents to preexisting structures) while distributing the cost of protection across private and public entities. The placement of homes in fire-prone landscapes appears to be an inescapable landscape reality and preordained future outcome. One of the only questions remaining, and left to fight over, is who will pay for and implement necessary mitigation activities.

PART IV After the Fire

THE CONCOMITANT EXPANSION
OF AFFLUENCE AND RISK

8 Dispatches from the Field

WIN–WIN OUTCOMES AND THE LIMITS
OF POST-WILDFIRE MITIGATION

By noon on Sunday conditions in the East Bay hills were bleak. From the base of the hills and looking east toward the origins of the fire above Highway 24, onlookers were confronted by a churning mass of swirling haze. The midday sun became increasingly obscured behind this rising and billowing column of charcoal, brown, and yellow smoke. Viewing the scene from afar, it would have been impossible to imagine that beneath this foreboding blanket of smoke a network of response units was busily crisscrossing the landscape fighting fire outbreaks and assisting evacuation efforts.

Real-time passages from audio recordings between first responders, dispatchers, central command, and battalion chiefs among others offer a visceral and unfiltered experience shedding light on the actual problems that accompanied efforts to respond to the largest urban wildfire in California's history. These insights serve as valuable learning moments for city managers, fire officials, and residents alike, as they provide verifiable evidence to support the implementation of appropriate postdisaster planning policy. As a review of the postfire capital improvement process in the second half of this chapter illustrates, however, the relationship between problem identification and management solution is hardly linear. Reconstruction efforts in the Tunnel Fire area have been mixed in their ability to address key problems that arose during the disaster response. Successful risk mitigation has largely followed the ability of proposed land and infrastructure modifications to deliver win–win financial outcomes for both the City of Oakland and its hillside residents.

DEPLETED WATER AVAILABILITY, NARROW ROADS, AND ACTIVE POWER LINES

Transcripts from field communications reveal three factors that inhibited firefighter capabilities during early stages of the firestorm: a rapid decrease in water pressure and availability; downed active power lines; and narrow roads blocked by parked and abandoned cars. To be sure, other factors such as radio frequency incompatibilities (discussed in earlier chapters) had an influence on fire response activities. However the three municipal infrastructure conditions identified here are shown to have severely impeded activities over and over during initial suppression and evacuation phases of the emergency response. This time frame marks a short window when the fire quickly expanded from a local brush fire to a full-blown wildfire. The rapid progression of official fire alarm declarations reflects the fire's swift movement into a vast swath of residential neighborhoods. In less than half an hour incident command progressed from a first-alarm to a sixth-alarm declaration. In fact at 11:26 A.M. and only twenty-eight minutes after calling a first alarm, the command team skipped the fifth alarm altogether, moving directly to sixth-alarm status.

Approximately one hour after the first alarm was ordered, the gravity of the situation led command team members to order the implementation of the hill area disaster plan. With such a short window of time to combat or escape onrushing flames, the emergence of water system failures, of active, downed power lines, and narrow, choked roads only exacerbated an already hazardous situation.

Audio records illustrate how the fire's rate of spread was overwhelming response teams within treacherous hill slope areas. At 11:31 A.M. a field operations unit placed an important message over the airwaves: "Chief, we've got the flames all the way up to Grizzly Peak, coming up the road . . .

about to cross over." The urgent tone marks an important moment in the fire's progression as it becomes clear that localized containment is no longer possible. The fire, now running ragged across hillslopes and up and down canyons, prompts an even more urgent message from Battalion 2.

OAKLAND 2 (DISPATCH): Command, we need air support *real* bad up here. They got anybody coming? [*emphasis in original audio*]

For many units on scene, the quickly progressing fire forced defensive strategies as opposed to more proactive mitigation strategies.

DIVISION A: Command, this is Division A.
BATTALION 2: Division A, go—you're probably not going to be able to stay there.
DIVISION A: It's coming over.
BATTALION 2: Correct, don't get anybody killed!

Another dramatic statement crossed the airwaves a few minutes later, the unnerving situation evidenced in Battalion 2's now urgent tone: "The fire's *totally* out of control on several fronts. We've got at least one hundred acres burning, trees, brush, houses. Give me five strike teams under mutual aid. Staging will be at Hiller and Tunnel." [*emphasis in original audio*]

Within fifteen minutes of initial reports indicating that flames were fast approaching the Grizzly Peak ridgeline (and county dividing line), operations confirmed the inevitable, albeit much more quickly than anyone expected.

OAKLAND 1: Let Contra Costa [County] know we are going to need their assistance. It's gone over the top of Grizzly Peak.

As the fire's rapid progression quickly challenged command teams (ultimately resulting in the first of many urgent calls for mutual aid), units on the ground were faced with untimely and dangerous infrastructure breakdowns. By late morning Engine 8 reached its position on a residential road to provide fire suppression and evacuation support to other units already on the scene. The path to assistance had been difficult and required navigating twisting roads, fleeing residents, and swirling smoke.

Finally in position at approximately 12:05, Engine 8 communication takes a noticeably distressed tone: "Command, this is Engine Eight!" Prompted by Command, the panicked voice continues: "We're running out of water up here! Is there anybody that could relay from the hydrant down at the bottom of Buckingham; we've got a four-way there. Anybody that can relay us the water? We need it bad."

Command responds, "See if you can find an engine company there; I'm sending 15 down that way."

Engine 8, now breathing heavy, exclaims: "Don't send anybody *down* Buckingham! There's parked cars here; the streets are completely blocked. Buckingham is blocked at the . . . Seven hundred block. We need somebody to come in at the bottom of Buckingham, pump into our four-way. We can't get down it. . . . There's power lines down!" [*emphasis in original audio*]

Later, and adding still further challenges to the already precarious situation, Engine 8 reports, "Emergency! Power lines down on hydrant!"

This final communication connoted one of the more desperate moments facing firefighters working the scene. At the very moment Engine 8 reported "power lines down on hydrant," a large hole instantly seared into the hose that was being used to extinguish incoming flames. Now faced with a rapid decrease in water pressure, the crew was forced to retreat inside a concrete block garage with other stranded residents. There they waited, unable to leave as the fire moved through the area directly outside. Fortunately both crew members and residents were able to escape unscathed after the fire front roared past.

Concerns over live power lines and reports of cars blocking access/evacuation routes can be found throughout the communication records. For example thirty minutes before Engine 8's passionate call for help, at 11:35 A.M. another engine on the scene exclaims, "Can you get PG&E [Pacific Gas and Electric] to shut down the power up here? We've got lines that are starting to drop all over the place, Chief."

As available water ebbed and approaching flames flowed, Engine 19 issues a highly panicked call to the battalion chief. The firefighter's voice is now at a shrill: "Battalion Two, we're abandoning the tank!"

Light on details yet heavy on expediency, Engine 19 exclaims less than a minute later, "We've abandoned!"

While Engine 19 was eventually able to move out of harm's way, others such as Engine 6 found it more difficult to evacuate the narrow winding roads and rush of smoke, heat, and whipping flames. For members of Engine 6, abandoning their position was too risky and they were forced to seek refuge in one of the only stable sources of water in the area: a nearby swimming pool. Submerged in the pool alongside the homeowner, responders spent more than an hour under the pool cover, periodically sticking their heads out to splash water on the cover to prevent its ignition. Ultimately the house burned along with all neighboring homes; yet the pool, its cover, and the three individuals beneath it survived.[1]

As if these already overwhelming conditions weren't enough, the news that no one wanted to hear was confirmed by Incident Command: "Engine Three, you're about to lose some houses on Golden Gate. And, um . . . the fire has jumped Highway Twenty-four." The fire, which had already burned clear through several entire neighborhoods, had bypassed the massive Highway 24 transportation corridor—a roughly five-hundred-foot swath of highway, off-ramps, and frontage roads that had been previously viewed as an insurmountable obstacle and firebreak. Indeed the City of Oakland's historical planning records indicate this concrete buffer had long symbolized a physical dividing line and historical marker between high- and low-risk fire areas.

Amid the expanding sea of smoke, sirens, and swirling embers many firefighters hastily hooked into hydrants only to find available water supplies reduced to a mere drip. This scenario played out again and again for response teams, particularly as the day wore on and new fire fronts emerged. As more firefighters arrived on the scene, the problem of inadequate water availability was only compounded. This confluence of misfortunes can be found in a short exchange between Command and an unknown engine on the scene: "Oakland Two, Emergency Traffic and Battalion Forty-four. We have houses on fire on Contra Costa. Houses on fire on Contra Costa." An unknown engine replies swiftly, "We've got a water shortage up here and we need some more."

Faced with new fire fronts and a lack of water, firefighters in the area continued to scramble to defend homes and assist residents along evacuation routes. Yet as the following communication shows, there were simply not enough resources available to cover the blistering firestorm. With an incredulous tone the command unit pleads for assistance: "I hope we've got *a lot* of help coming, John [Baker], because this thing is now going towards Broadway Terrace—I guess you heard they had an outbreak. They've got houses going on Contra Costa. We've got multiple new fronts."

Chief Honeycutt interjects, "Chief Matthews, Chief Honeycutt."

Command prompts, "Go ahead, Neil [Honeycutt]."

Chief Honeycutt, having assessed the scene, implores Command for more hands on deck: "I'm at Contra Costa and Buena Vista. I've got about fifteen houses involved. Helicopter 106 is working our flank. I've ordered ten additional engine companies. If we can move an engine company from the top of the hill that's not committed down here, we need them desperately."

Command replies swiftly: "Top of the hill are *totally* committed and overwhelmed, Neil."

Tragically, amid this frantic response effort, two city emergency responders lost their lives in the fire: Oakland Division A battalion chief James Riley and Oakland police officer John Grubensky. The aforementioned problems of active, downed power lines and narrow, vehicle-choked roadways afflicted both Battalion Chief Riley and Officer Grubensky. According to Lamont Ewell, who was Oakland fire chief at the time of the fire, Riley and Grubensky were found with the remains of those they were trying to assist.[2] Battalion Chief Riley was helping a resident flee the fire when it is believed that a fallen power line electrocuted both him and his evacuee. (One of Chief Riley's last, ominous communications is provided in a passage near the beginning of this chapter. He appears as "Division A" and is told not to "get anybody killed." He reported his final contact just a few minutes later at 11:44 A.M.) The body of Oakland police officer John Grubensky was found along with the bodies of five civilians stuck behind a collection of jammed vehicles at one of many narrow, winding road segments in the steep, hillside fire area. Trapped at this bottleneck, Grubensky and others were unable to escape the rush of smoke and fast-advancing flames. His body was found lying on top of a woman he was helping to evacuate. In both cases fatalities occurred not because of poor emergency response decision making but rather because hillside infrastructure impeded vehicle evacuations and the eleventh-hour heroism of these brave civil servants.³

OVERCOMING THE ARTIFACTS OF PLANNING POLICY:
THE SEARCH FOR WIN–WIN OUTCOMES

The above emergency radio communications provide useful insights into three primary infrastructure categories that hampered response efforts: water supply, roadway impediments, and power lines. These problems and resulting efforts to troubleshoot them can be found within a wide range of expert panel commentaries and blue ribbon reports developed after the firestorm. Recommendations ranged from new interagency communication protocols to modified vegetation management and home construction practices. In many ways the Tunnel Fire has served as a crucial catalyst for the implementation of significant local fire mitigation activities around the entire West. New fees, education programs, enforcement protocols, and governance arrangements all emerged in the months and years after the fire. Regionally and nationally the Tunnel Fire (or "Oakland Hills Firestorm" or "East Bay Hills Firestorm" depending on who is reporting) remains *the* urban wildfire reference point in U.S. history. (Others such as the Chicago and San Francisco fires are considered strictly urban fires.) Due to its notoriety as a "worst-case scenario" wildfire the devastating Tunnel Fire sparked substantial changes in fire departments and agencies across the United States.

In Oakland these efforts have seen mixed success. Interviews with city officials and a close reading of city documents suggest that two factors most often influence efforts to implement substantive risk reductions: private property interests and city budget considerations. These factors have negated progress in some areas while facilitating significant risk reduction advancements in others. Where progress is achieved it is because mutually beneficial outcomes are identified for homeowners and the city. In nearly all cases progress has been influenced by the ability of capital improvement recommendations to retain if not enhance the quality of life and levels of affluence for hillside residents. This is how postdisaster reconstruction proceeds: through the actuation (intentional or otherwise) of win–win outcomes that deliver tangible financial benefits to residents and cities alike.

When Risk Constructions Get Instantiated in Water Infrastructure

Many firefighters experienced water supply challenges just as fire activity was increasing and the need for water was critical. Several factors led to this untimely development. First, water capacity infrastructure in the area remains a legacy of historical pipe installations. In the Rockridge district for example, which comprises much of the area consumed by the Tunnel Fire, water infrastructure was put in place between 1910 and 1935. Four- and six-inch mains that are widely considered insufficient by today's municipal standards anchored this pipe infrastructure. According to an expert panel convened in 1981, this area was deemed highly developed and urban and therefore of lower fire risk than areas upslope. Because of this historical fire designation portions of the Rockridge area were never prioritized for water pipe system updates.[4] Furthermore city officials viewed the Highway 13 complex as a sufficient firebreak that would decrease fire activity in the downslope area. Generally speaking, areas on the more heavily open-space (east) side of the freeway were deemed high risk, while neighborhoods on the city (west) side were considered lower fire risk (see Chapter 1). This social construction of risk across the landscape wound up putting response crews at a great disadvantage. As a consequence many locations simply lacked adequate water pressure to apply to burning vegetation and structures.[5] Most hydrants in areas deemed to be low risk were limited to a maximum supply of roughly 500–650 gallons per minute (gpm). On the other hand, high-risk zones received more recent upgrades that increased pressure to 1,000 gpm or greater.

This was a well-recognized problem for the Oakland Fire Department. One year prior to the Tunnel Fire a resident in the fire zone noticed a nearby home structure on fire. Firefighters were on the scene putting out the blaze. Noticing what appeared to be an extremely protracted struggle to subdue the flames, this individual described asking one of the firefighters about the persistent fire. According to this community member, the firefighter explained:

Well you know we're doing what we can, but this particular section, the water pressure's really low here. So we just can't throw enough water on their house and it's frustrating for us, but we're doing what we can do; but it's burning quicker than we can put it out.

Several years after the fire the East Bay Municipal Utility District (EBMUD) and the City of Oakland joined in a cost-share project to install new water pipes in this portion of the Tunnel Fire area. The Rockridge Area Water System Improvements Project increased water flow to all hydrants in the Rockridge Special Assessment District (RSAD). Along with payments from EBMUD and the City of Oakland, 750 parcel owners in the RSAD pay $135 a year via county tax rolls to cover the incremental costs of pipe upgrades, which increased improvements from a proposed 1,000 gpm level to a desired performance of 1,500 gpm.[6] The role of outspoken citizens was crucial to this process. Their vocal participation in postdisaster planning activities and their willingness to pay extra for piping system upgrades were critical to overall project success. Replacing water infrastructure is typically a very expensive and onerous undertaking. However, because roads and underground utilities were already being replaced, residents seized upon this unique moment in time. Recognizing this opportunity, neighbors banded together and pressured the city to maximize infrastructure upgrades. Of course they also knew that modernizing pipes to the 1,500 gpm level would enhance overall neighborhood quality and contribute to its fast-climbing real estate values.

HYDRANT NOZZLE CONNECTIONS AND THE TECHNOLOGICAL
LIMITATIONS OF MUTUAL AID

A second water infrastructure problem during the firestorm involved fire hydrant nozzle connections. Mutual aid has always been recognized as an important support structure in the context of major disasters like the Tunnel Fire. Increasingly fire and other emergency response department are coming to rely on outside support as state rollbacks and budget restrictions make autonomous city disaster governance ever more difficult. And yet this "flexible" governance structure has revealed its own limitations: many responding units from outside areas were unable to connect to Oakland hydrants due to incompatible nozzle hookups. Most of these units only discovered the connection issue after entering the critical combat zone. Although suitable hookups were available along lower portions of some hydrants, coordinating agencies and participating firefighters were not sufficiently informed of these contingency features. Ultimately these units were left scrambling for suitable nozzle adapters instead of trying to extinguish the surrounding fire.

When California adopted a standard two-and-a-half-inch threaded connection for hydrants across the state, Oakland and San Francisco opted instead—in a cost-saving measure—to maintain their three-inch connections. As a precaution they kept a supply of adapters on hand for mutual aid units. The plan required that adapters be obtained from warehouses to meet incoming strike teams at staging areas. Because the Tunnel Fire occurred on a Sunday, efforts to retrieve the adapters out of storage and distribute them to on-scene supply trucks were delayed.

In the years following the Tunnel Fire the City of Oakland sought to reduce its municipal exceptionalism and increase its regional interoperability. By the end of 1998 the city had changed over six thousand hydrants to two-and-a-half-inch national standard thread connections.[7] In the efficient cost-sharing environment that marks contemporary neoliberal hazard management governance, the issue of hydrant adapters (as well as incompatible radio frequencies) should serve as a cautionary tale. As fire mitigation partnerships and coalitions continuously shift and evolve, future agency asymmetries—in both knowledge and technology—will surely emerge. These incompatibilities should be vetted early and prior to emergency collaboration.

FIRE, POWER, AND THE CHALLENGE OF COUNTERING GRAVITY

The hill area water conveyance system—like other foothill and mountainous areas around California and the West—is designed as a layered network of pressure systems where water from lower elevations is pumped into reservoirs located uphill. Water contained in these reservoirs is then delivered downslope by gravity to street hydrants and individual water lines. In order to retain adequate water volume in reservoirs, these systems depend on electrically powered pumps to deliver water back upslope. This system is designed to resupply reservoirs during high-demand periods such as fire events. Loss of electricity to these crucial water pumps presented a third water-related problem for firefighters just as emergency response activities required full flow capacity.

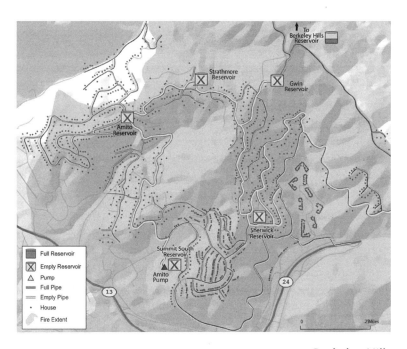
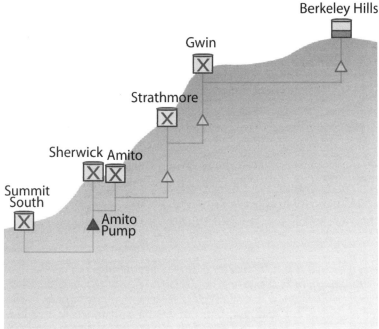

Map 7. The Hiller Highlands area and origin of the Tunnel Fire (top). Due to power line damage, the Amito Pump lost power at 11:35 A.M. Without new water pumped to upslope reservoirs (bottom) nearly the whole region depicted ran out of water by midafternoon, with the first reservoir running dry at approximately 1:50 P.M. (Credit: Allie Hausladen and Nicole Brunner)

Throughout the hill area, several pumps lost power as a result of damaged electricity lines. By early afternoon nonoperating pumps left many hill reservoirs and the neighborhoods that depend on them with an insufficient supply of water. For example power losses from falling power lines caused the Amito Pump located at the base of the burning Hiller Highlands neighborhood to lose power at 11:35 A.M. As a result firefighters in pressure zones located above this pumping station (see Map 7) were left only with water already in the reservoirs until emergency pumps and generators were brought in at 6:00 P.M.

In response to these power system failures EBMUD acquired portable pumping units for deployment during emergency events. All pumping plants are now equipped with an emergency generator connection and some of the plants have an onsite emergency backup generator.[8] These mark important improvements that will enhance fire preparedness in the future. In hindsight however the lack of adequate contingency infrastructure support at the time of the Tunnel Fire remains an alarming development. Yet when viewed in the context of a long list of response difficulties, the inability to resupply power to vital water pumps becomes somewhat less surprising. Having undergone major budget restructuring in response to California's shifting taxation and city investment environment during the latter half of the twentieth century, the City of Oakland and its various municipal service departments were left ill prepared for a disaster of this size and magnitude. (This concern rests on top of much more foundational questions about why homes were ever constructed—in fire-prone landscapes and with a tenuous, technology-dependent supply of water—in the first place.) In summary, while each of the response team communications reported above lasted only a few minutes, the factors causing these mitigation barriers and system failures were decades in the making.

Enabling Evacuation: Increasing Vehicle Throughput by Reducing Road Friction

Water is generally a compliant and malleable substance. Water flowing through narrowing rivers for example will increase velocity to retain a consistent volumetric flow within the entire river channel. Due to the fluid property of water, increased velocity in narrow channels preserves the principle of continuity, enabling a steady and constant throughput of water despite irregular channel size. This flow rate increase is important for it prevents waterways from overflowing at the head of contracting channels. Water thus naturally resists the formation of bottlenecks.

Unfortunately cars do not behave like water. When eleven individuals perished in a traffic jam on Charing Cross Road it was a harsh reminder of these problematic physics. This deadly bottleneck occurred during a short peak evacuation period when fleeing residents and emergency vehicles converged along a narrow segment of the roadway. Stuck behind entangled vehicles, some individuals left their cars to flee on foot. These abandoned vehicles further added to the gridlock. Unable to navigate the blockade, trapped individuals were quickly engulfed by the fast-moving smoke and flames.[9]

Of course the inability of these eleven individuals to escape should not be blamed on the low viscosity of three-thousand-pound passenger vehicles. Many roads in the area, including portions of Charing Cross, are only *half* as wide as minimum national standard requirements. Most of these narrow roads, measuring twelve to fourteen feet wide, were constructed in the 1920s. Unfortunately they were not brought up to modern road-width standards when more homes were constructed in the 1970s and 1980s—this despite recognition of high fire risk dangers and the need for robust and effective evacuation capacity. If there was any silver lining to the Tunnel Fire it was that the postdisaster reconstruction process would present a golden opportunity to increase road widths and help reduce the possibility of future bottlenecks, while simultaneously aiding the movement of response units into the fire area. There was widespread agreement within the city about the need for these road improvements. Suggestions included prohibiting parking on all roads under a minimum width, designating one-way roads, widening hairpin turns, establishing cut-bank parking spaces, and widening narrow roadways where possible (see Figure 19).

The problem of vehicle bottlenecks along narrow roads was compounded by the presence of parked cars, which in some locations constricted vehicle travel to a single lane. Parked cars did not just impede rescue and evacuation routes. As flames swept across roadways, these highly combustible vehicles wound up contributing to the area's overall

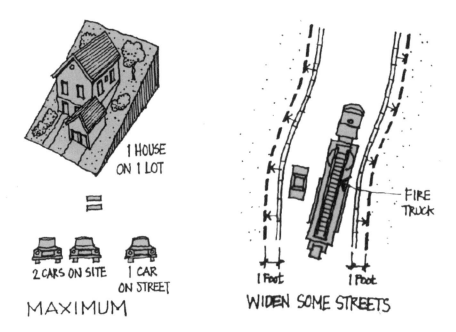

Figure 19. The City of Oakland produced many reports to tackle the problem of narrow, blocked roads in the hills area. A planning document depicts proposed road and parking changes in the East Bay hills. (Source: 1992 California Emergency Design Assistance Team; CEDAT Guide)

fuel load and combustion potential. Moreover, as can be seen in Figure 20, numerous cars cluttering streets have extensive damage to their rooftop. As it turns out, many of these local vehicles were not originally left along roadways. Instead these cars were suspended in elevated home garages directly upslope. As garage foundations burned and crumbled, vehicles fell through their weakened floors, ultimately tumbling downhill like boulders until they came to rest along flat road surfaces. While city planners did not directly address the rolling-car phenomenon, the issue of street parking was a major focus of postdisaster hazard mitigation efforts.

GRIDLOCK: WHERE PROPERTY INTERESTS, BUDGET CONSTRAINTS, AND ROAD IMPROVEMENTS CONVERGE

Several suggestions were offered to limit parking in the fire area. These included the enforcement of no-parking zones (particularly along narrow

Figure 20. A vehicle bottleneck along one of several evacuation routes. Vehicles became stuck while going in different directions and encountering parked cars along a narrow road. Downed power lines can be seen in the background.

street sections); the use of bollards along sidewalk edges; establishing a maximum allowable number of vehicles per private property; encouraging off-street parking considerations in new home rebuilds; and the development of alternative transportation options in the area. Unfortunately concerns over private property values, budget restrictions, and local topography rendered the vast majority of these proposed road modifications intractable.

A first obstacle hindering road modifications was the sheer number of parcels required for expanded right-of-ways—an area comprising scores of homeowners loath to concede a portion of their property to the city. A few feet of increased egress along a one- or two-hundred-foot property line will result in several hundred square feet of property loss. This stoked fear among homeowners about the negative impact such a concession might have on property values. Most homeowners also deemed parking limitations unreasonable. Would homeowners have to sell their vehicles if unable to comply with strict city restrictions? Would the city have to draw enforcement resources from other city programs to issue parking permits in a region with comparatively low home densities? Local topography presented yet another challenge due to the expensive nature of street cutbacks. Increasing road width in level neighborhoods is one thing. But enacting such alterations in areas with steep hillslopes, canyons, and retaining-wall requirements is considerably more costly and time consuming. Further complicating matters was the fact that these substantial road improvement costs were (and still are) proposed during a period already containing an immense backlog of road construction projects. For example the 2012–13 budget for street renovations in Oakland was $4.3 million. This is compared to the current backlog of street renovations, which sits at $435 million.

The implications of these challenges are clear throughout the fire area. Consider the case of Charing Cross Road and the site of the fatal traffic jam that took the life of Chief Riley. Here private property value concerns and feasibility limitations due to topographic challenges and already stretched budgets have led to very minor modifications. In a few locations the road has been slightly widened. But these modifications are not significant enough to substantively alter access capacities in the area. According to one retrospective report, "The lesson here is to resist making concessions on initial development patterns, lot configurations, road alignments or infrastructure standards ... access, lot size and the footprint of development cast the die for every community."[10]

Power for the People: Utility Line Hazards during the Tunnel Fire

Some of the most striking accounts emerging after the fire involve the dangerous presence of ignited and downed power lines within the impacted landscape. It is easy to forget that throughout many fire-prone areas electricity zips overhead, down roadways, across canyons, and up ridgelines. Power flowing through these communities is at once a crucial source of energy for home and community services, and a potentially deadly conduit of high-voltage electrical currents—"overhead ribbons of explosive fire," as a retrospective FEMA report referred to them. The occasional fork in toaster or finger in socket aside, electrical currents and city residents typically remain safely separated. But these secure conditions become disrupted in the presence of large fires, which can raise temperatures in actively burning areas to well over one thousand degrees Fahrenheit.

By late morning on October 20 many streets were already blocked by burning utility poles and live power lines. Meanwhile flying embers from exploding transformers swept across the sky landing on homes and vegetation. The dangers of exposed electrical currents for people moving along roadways were significant and resulted in injuries, fatalities, blocked evacuation routes, and impeded firefighting capabilities. Moreover as power lines severed, water pumps failed. Tanks sitting upslope that relied on pump-driven water conveyance quickly began to run out of water. Once high-voltage lines shorted out at 1:15 P.M., power to the remaining pumps ceased as well. In the following hours a large portion of the water system in the hills began to run dry.

Power line arcing at an adjacent electricity substation contributed to the fire's spread. During a fire event power line arcing occurs as extremely hot air passes above high-voltage lines due to the thermal updraft created by the fire. The arcing phenomenon caused power line segments—which extend from the substation for several miles south into the hill area—to ignite like a necklace of high-intensity, glowing orbs suspended above the ground. Radiating beneath a thick layer of smoke and extending across grass and forested landscapes, the arcing process generated an extensive shower of sparks producing new grass fires that eventually merged with the main body of fire.[11]

POWER TO THE PEOPLE: UPGRADING UTILITY LINES
IN THE FIRE AREA

The magnitude and severity of these potential power line failures seemed to leave a strong impression with many officials. As one postfire report noted, "On many occasions electricity suddenly disappeared: power poles had burned down, breaking the lines; or the fierce winds had whipped and broken other lines; or huge clouds of smoke and gases had enshrouded the lines, causing them to arc and trip relays." This quote conveys more than a sense of astonishment however. It also indicates recognition that the problems power lines present for evacuation and emergency response efforts will require greater scrutiny moving forward. And while the 1991 firestorm certainly illustrated these problems, this quote actually stems from a September 22, 1970, Oakland Hills fire report—a fire that took place in nearly the same location as the Tunnel Fire. For at least twenty years then city officials were well aware of the extremely dangerous conditions created by live power lines. And yet very little was done to rectify the dangerous problem.

The 1991 fire however presented a sufficient tabula rasa—with over three thousand homes and countless power poles destroyed—to actually engage in landscape-scale electricity infrastructure modifications. But the path to power line upgrades was not so straightforward. Line modifications were limited to areas impacted by the fire despite the whole region's susceptibility to these problems. In fact due to budget limitations infrastructure was only placed underground in areas experiencing sufficient damage to utility infrastructure. Given this spotty treatment, as one leaves the fire perimeter, power poles reappear along roadways. And in certain areas within the fire perimeter overhead lines still exist. This means that during the next major fire electricity lines will still pose a significant threat to local residents and emergency responders—this despite more than sufficient awareness of both the problems and the remedies at hand.

The destroyed power systems that were modernized and placed underground emerged largely as a result of the actions of community members who had the time, money, and resources to ensure a form of redevelopment that went beyond mere infrastructure replacement. Initially power companies intended to install aboveground poles, transformers, and lines that were similar to preexisting capital. Because money was not available for further upgrades, residents in many of the affected areas responded aggressively by conducting research and speaking with various politicians and key members of the planning community. Ultimately residents agreed to pay one-third of the cost of infrastructure improvements.

Although actual payment responsibilities shifted over time, this process illuminates the crucial role of affluent neighborhood activists in shaping postdisaster development outcomes. Leveraging various social advantages (such as having ready access to influential city officials), these communities were able to garner substantial control over planning decisions. Without question financial incentives were front and center in this process. As a resident and chief community negotiator mentioned, the infrastructure upgrades "would improve the value of our homes. Let's get real here, it was good for us."

Reported in the first part of the chapter, "dispatches from the fire" have helped to identify challenges experienced by residents and responders during the hazard event itself. These real-time emergency communications in turn allow us to match event response problems with substantive management actions. Based on these transcripts, several important issues emerge. A survey of postfire water, road, and power infrastructure reconstruction efforts illustrates that progress toward reconciling these issues has been mixed. Capital improvements were driven largely by private property considerations and their commitment to leverage the disaster in pursuit of neighborhood enhancements and estate-based wealth accumulation. Upgrades to water and power line equipment were lobbied and partially paid for by determined residents who used their positions of privilege to engage in collectivized risk reduction. In short, in cases where potential neighborhood and property value improvements were apparent, the community was willing *and able* to supplement beleaguered city budget capacities and help pay for municipal upgrades. The fire thus presented an opportunity for residents to secure financial enhancements over the long run. This presented a win–win for residents and the city alike. However when private benefits were less evident (or simply not attainable)—as was the case with road-widening initiatives—residents were less apt to financially or logistically support such recovery efforts, and the pursuit of win–win outcomes unraveled.

9 Out of the Ashes

THE RISE OF DISASTER CAPITALISM
AND FINANCIAL OPPORTUNISM

In the years following the fire, weary community members slowly returned and resettled the impacted area. For both new and old residents the once desolate landscape gradually came to look and feel like home again. New trees and shrubs were planted to replace fire-damaged vegetation. Roads were eventually repaved, mailboxes lined roadways, and neighbors could again be found roaming the sidewalks. It took considerable time and effort but nearly a decade later most of the homes in the area were rebuilt. And while the old neighborhood feel—with historic wood structures set against soaring tree lines—was no longer, a new neighborhood and sense of community had emerged. The rebuilding process was also one of neighborhood rehabilitation and renewal, marking a profound revival of place. Like the Greek mythological figure of the Phoenix, which exists through a cyclical pattern of birth and rebirth, the Tunnel Fire area community had risen from out of its own ashes, indefatigable in spirit, resilient, and determined to recreate life and home once again.

The symbol of the Phoenix was utilized by many in the area as a reminder of their own difficult yet resolute journey. Groups such as the North Hills Phoenix Association (later merging to become the North Hills

Figure 21. A sign placed on a lone eucalyptus tree by a community member who lost her home just days earlier to the Tunnel Fire. Out of the fire's ashes rose a renewed and restored sense of home and community. Also emerging were new real estate and private sector profit-making opportunities.

Community Association) leveraged this spirit of rebirth in their community name and vision. A resident whose home had been destroyed hung a three-by-five-foot sign on a large eucalyptus tree in the middle of a five-way intersection (the same tree referenced in the Introduction) within days of the fire. The handwritten sign borrowed themes from the Phoenix, reading, "May we all be restored and renewed." For the next several years cars and trucks, residents, and construction workers all passed by the increasingly tattered and dusty sign, reminded of the resilient nature of the neighborhood and its residents (see Figure 21).

But determination and persistence were not the only qualities to emerge from the fire's ashes. Also materializing from the rubble was a clear and working sense of financial opportunism. These diverse and strategic efforts to extract profit from the postdisaster landscape raise questions about what precisely was being restored and who exactly was being renewed.

HOME AUGMENTATION AND ESTATE-BASED OPPORTUNISM

Naomi Klein first introduced the theory of "disaster capitalism" in her best-selling book *Shock Doctrine* in 2007. Here Klein suggests that postdisaster landscapes—laid bare both physically and politically—present opportunities for new economic reforms, transformations, and material inscriptions that could only ever be achieved in the wake of such upheaval. From new tourist economies along post-tsunami coastlines in Southeast Asia to the privatization of the "war on terror" in destabilized, post-9/11 regions of the Middle East, Klein shows how free market policies have advanced through the exploitation of disaster-shocked peoples and landscapes. Theories of disaster capitalism are applicable—on a much smaller scale—to the Oakland Hills postfire area (as well as other fire sites around the West). Many residents describe how solitary chimneys and ashen rubble stretching for miles in each direction made the hill area look "as if a nuclear bomb had been dropped." While this may be an evocative if somewhat exaggerated description, the stripped and desolate postfire landscape has also come to represent another important baseline: a tabula rasa upon which new market transactions, private property interests, and estate-based wealth opportunities emerged.

Lot by lot new homes slowly repopulated the fire area in the months and years following the conflagration. The redevelopment process was anything but subtle as construction teams shifted from one site to another across the hillside. Like a caterpillar procession in search of new pupation sites, service trucks big and small edged slowly along a network of undulating roadways in search of properties to excavate, frame, drywall, roof, and paint. With over three thousand lots in need of reconstruction this was a productive several years for the residential architecture and home construction industries. But perhaps more importantly the postfire environment provided a brief window for strategic capital investment and financial opportunism for private property owners. This investment did not merely represent a short-term spike in services, business activities, and profits as witnessed by the construction sector. Rather this transformation marked an increase in property *value*, hence symbolizing a recalibration and leap forward in net estate worth for those holding a staked claim in the hill area. This process is linked to (and indeed substantiates) what Mike Davis posits as "upward social succession"—an increase in land development and property values that follows natural disasters. This is a process that ensures subsequent residents will occupy a class status similar to or greater than the current population's.[1]

The broad historical sweep of hillside subdivisions and home developments stretching back to the late 1800s established a foundation for future capital accumulation and elevated postdisaster property values. After the Tunnel Fire displaced residents worked with insurance companies, city construction review boards, and architects (by no means an easy gauntlet to navigate) to negotiate new home construction details. Analysis of home footprints before and after the fire indicates a remarkable transformation: houses became larger in size and closer to one another. Geospatial analysis was applied to 422 home footprints before and after the 1991 fire. Study results reveal that among these structures average home size increased by 247 square feet (11.1% larger than the previous structure). The distance between each home was reduced from 14.2 feet to 12.4 feet (or 14.4% closer). Figures 22 and 23 present a sample depiction of changes in home size and proximity after the Tunnel Fire.

What did this expansion actually look like at the level of the home? And how did it influence property values? While a comprehensive list of changes would be difficult to attain from historical records, census data were analyzed and averaged across tracts heavily impacted by the 1991 fire to determine changes in the number of bedrooms of newly constructed homes. These findings were compared with similar home analysis in non-affected tracts. Results indicate that the number of bedrooms in homes within impacted areas increased between 1990 and 2000 by 0.15 bedroom per home on average in the decade after the fire. Homes outside of the affected areas remained the same. Meanwhile the value of homes within these impacted tracts increased at a rate that is nearly double the home value increases in other areas throughout the city (Figure 24).

Building Back Bigger: The Case of Suburban Colorado

A similar study of home size in Colorado reveals comparable results.[2] The 2012 Waldo Canyon Fire burned over 28,000 square miles at the

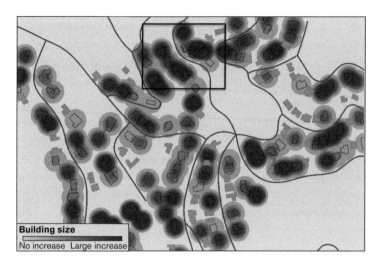

Figure 22. Kernel density analysis depicting homes in the Tunnel Fire area (from larger sample of 422 houses). Image shows growth in home footprint after the rebuild. Darker shading indicates greater increase. The footprints of rebuilt homes were on average 247 sq ft (or 11.1%) larger than prefire structures.

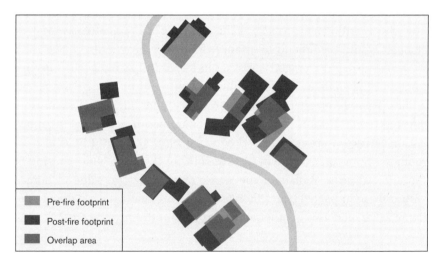

Figure 23. A subset of houses (box inset in Figure 22) depicting an overlay of pre- and postfire home footprints in the Tunnel Fire area. Nearest-neighbor distance analysis was applied to detect a change in the proximity of homes. Postfire structures were on average 1.8 feet (or 14.4%) closer than original homes.

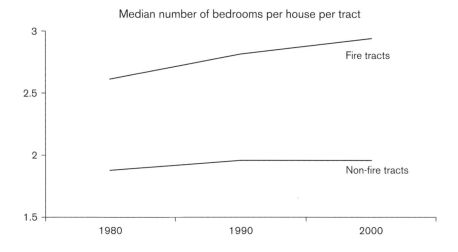

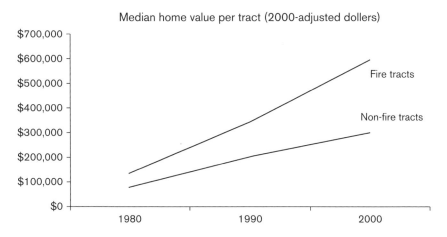

Figure 24. Top: Number of bedrooms in homes increased between 1990 and 2000 reflecting the interim rebuild process. Homes in census tracts with homes burned in the Tunnel Fire added approximately 0.15 bedrooms per home. *Bottom:* The value of these same homes increased at a faster rate than all other tracts averaged over the same period.

wildland-urban interface just four miles northwest of Colorado Springs. At the time it was the largest fire in Colorado history in terms of homes destroyed—only to be surpassed by the Black Forest Fire a year later. With its origins in the mountainous Pike National Forest the Waldo Canyon Fire spread in several directions including east into the Colorado Springs suburban Mountain Shadows neighborhood. Comprising large ranch and two-story homes, the Mountain Shadows community was constructed in the 1980s–1990s along a belt of rolling open grasslands and pine forest rising upward through rugged terrain into the Rocky Mountains. As the fire encroached on this foothill community—first ignited by spot fires well in advance of the main front—furious winds whipped several single-home ignitions into a firestorm of grand proportions. By the time the fire was finally extinguished, 384 homes in the Mountain Shadows neighborhood had experienced either full or partial burn damage.

As of 2014, 248 of these homes had been rebuilt with a surprisingly similar outcome to the Oakland Hills Tunnel Fire area. Across all rebuilds home size increased by 14 percent (from 3,230 to 3,676 sq ft, or 446 sq ft bigger). While it's difficult to know precisely where this extra square footage was applied, real estate data show that there were on average 3.67 bedrooms per original home and 3.85 bedrooms in rebuilt homes (an increase of 0.19 bedrooms). Targeting the number of bathrooms was apparently an even more popular growth strategy. The average number of bathrooms before the fire was 2.82 per home, rising to 3.15 after the fire (an increase of 0.33 bathrooms). Ultimately the home reconstruction process in both Oakland and Colorado Springs points to a noteworthy if not surprising trend: the construction of homes that are appreciably larger than before.

Everyone Wins? The Rush to Get Homes and Tax Revenues Back on the Landscape

Several factors influenced this reconstruction trend in Oakland (and the rebuild process in Colorado and elsewhere). The first factor involved construction-permitting procedures, which were hastily developed immediately after the fire, leading to inadequate oversight of rebuilt structures. According to one member of the design review panel in Oakland, only after the rush for reconstruction had settled down and further precision was applied to the permitting process did the pursuit of a coherent and consistent neighborhood feel begin to emerge. This shift involved giving greater scrutiny to the aesthetic cohesion of rebuilt structures and included promoting consistent space and volume allocations within newly constructed homes (though the process of defining "aesthetic cohesion" was itself highly contentious). Considerably less attention was given to scrutinizing the overall square footage of homes. While particular home geometries and styles captured the attention of review board members, overall home volume appears to have received far less attention.

Second, there was a general desire by the City of Oakland to accelerate the return of households to their hillside properties. After all, thousands of residents suddenly found themselves thrust into new and unfamiliar living arrangements with much longer commutes (to school and work) and a disrupted sense of community belonging. In order to assist thousands of fire victims in the area an expedited design review was promoted that would fast-track permits for approval. This pursuit only further enabled an environment of lax oversight in the fire's aftermath.

Third and perhaps most importantly, fast-tracking reconstruction was an effective strategy for cities like Oakland and Colorado Springs to recapture an important tax revenue stream for its many municipal services. While short-term state and federal grants were made available to cities and homeowners to compensate for lost property tax revenues, the uncertain duration and permanence of these supplemental payment streams motivated city officials to expedite the rapid return of homes to the landscape.

The road to riches was not so simple for the City of Oakland however. While home values increased significantly after the fire, corresponding tax revenue increases were slower to emerge as a result of legal provisions under Proposition 13, which prevents substantial rises in property taxes for homes until a transfer in ownership. Yet there was some prospect for increased revenues for the city as the Tunnel Fire led many homeowners to simply never return. This meant new residents bought into the area, which activated a jump in property tax revenues—a trend that continues to this day as older residents who purchased homes during the 1970s for under $100,000 are now selling properties at price points well over $1,000,000.

Oakland is particularly dependent on these revenue-generating property taxes. It is a VHR (very high risk)-landscape city containing perhaps the most combustible tax base profile in the region—drawing nearly $110 million in property tax revenue from these areas in 2012 alone. This tax profile becomes a *risk profile* when revenues are placed in areas increasingly susceptible to firestorms. For the City of Oakland the hillside (including the Tunnel Fire area) presents a crucial yet risky financial asset, a reservoir of annual tax revenues that can be depleted (at least temporarily) with the strike of a match. It was precisely this lost revenue that contributed to the hasty return of homes to the hillside.

Of course for homeowners the logic for building back bigger was simple: capitalize on the disaster and generate as much square footage (and property value) out of insurance claims as possible. As we see from assessed homes in Colorado and California, part of this strategy included adding an extra bedroom (and for many observed Colorado homes an extra bathroom) in order to increase home values. When entire neighborhoods undergo this transformation a multiplier effect takes hold, for the property values of all homes increase alongside their neighbors' and well beyond individual capital investment levels. The tide of single-estate insurance monies lifts all neighborhood boats. Meanwhile the firestorm and its resulting pattern of home augmentation presented financial benefits for others involved in the postdisaster real estate market. For example larger and more valuable homes injected more money into local real estate market transactions resulting in higher agent commissions.

In the wake of the largest firestorm in California's history, two trends from the reconstruction process emerge. First, various groups and individuals profited off the newly rebuilt landscape in a manner consistent with Naomi Klein's *Disaster Capitalism*. Second, increased overall fuel load in the form of larger homes and more proximate structures have arguably generated *higher* susceptibility to future fires. (While it is true that many homes now incorporate more stringent fire-safe standards, they also add more overall fuel load to the landscape and if ignited pose a greater risk to their now closer neighbors.) Through the postfire rebuild process we see the simultaneous deepening of both fire risk and affluence in the Tunnel Fire area.

FIRE FOAM, SNAKE OIL, AND THE PRIVATIZATION OF FIREFIGHTING

American firefighters are iconic figures that have come to symbolize the honorable, selfless, and hardworking actions of public servants around the United States. After September 11, 2001, firefighters came to represent even more: the indefatigable and unifying spirit of American patriotism. While their status as honorable public servants certainly continues, over the past several decades the firefighting community has come to reflect an American ethos of a markedly different kind, one that actually challenges the integrity, scope, and composition of the public firefighting sector itself. As a global pioneer of economic liberalism the United States remains at the forefront of efforts to promote fiscal austerity and privatize the public sector. If we look closely we will see the collision of these two American brands: economic freedoms embedded in the private market, and the patriotic spirit of the American firefighting industry. What has emerged is an industry landscape composed of rapidly changing firefighter roles, actions, and priorities. Areas within the WUI are serving as ground zero for this new, modern American firefighting economy.

Similar to private contractors like Blackwater operating under the shadow of the U.S. Armed Forces, a number of private firefighting and prevention enterprises now work alongside the traditional structure of city, county, state, and federal fire agencies. Across the country private companies are increasing in numbers. In 2012 there were 256 private firefighting companies in the United States—a number industry forecasters expect will grow to more than 320 by 2017. Over the same period the number of private firefighters is expected to increase from 16,880 to 27,200. As the website of a leading community fire information portal put it, although private firefighters "make up just 4.3% of the nation's total firefighters . . . this is an industry on the verge of catching fire because of growing trend towards privatization."[3] As this sector "catches fire," the term *fire engine company* takes on a whole new meaning.

For homeowners who can afford it, use of companies like Firebreak Spray Systems (FSS) is a logical choice. Firebreak Spray Systems is a private firefighting business that has partnered with the insurance company American International Group (AIG) as part of its Wildfire Protection

Unit. Under this arrangement homeowners pay AIG extra for high-end insurance and in return receive concierge-level firefighting services from FSS. Upon early fire notification small firefighting teams enter at-risk neighborhoods and spray houses with fire retardants to protect against incoming flames and flying embers. For those willing to pay premium rates the benefits of private fire prevention appear straightforward. As a resident in Malibu, California, and a client of Firebreak commented: "Here you are in that raging wildfire. Smoke everywhere. Flames everywhere. Plumes of smoke coming up over the hills. Here's a couple guys showing up in what looks like a fire truck who are experts trained in fighting wildfire and they're there specifically to protect your home . . . It was really, really comforting."[4]

Private wildfire protection is increasingly popular in other states as well. For example a number of private services are available in Colorado where three massive fires destroyed hundreds of suburban homes between 2012 and 2015. Here the Chubb Group of Insurance Companies contracts with Wildfire Defense Systems, a Montana company with ready access to trained fire professionals and fire engines. For insurance companies like Chubb and AIG this marks a savvy approach to minimize settlement costs by actively reducing the number of homes burned. As FSS chief executive Jim Aamodt notes: "We are saving homes that may average $3 million to $5 million. . . . Those are the hard dollars, money the insurance company is not paying out."[5]

A Hard Sell: "Water on Steroids" and Its False Sense of Security

Not everyone is so quick to accept the participation of private companies in active fire settings. Criticisms fall within two broad categories: the first area involves health and safety concerns, while the second consists of broader ethical considerations. According to a high-level fire official in Northern California, homeowners should view private fire companies with skepticism because residents oftentimes believe they will get a level of enhanced protection simply by spending extra money. She notes that part of the problem is the illusion of security generated when "these Type 6, you know, firefighter lookalikes . . . [are] running in there with foam and providing this concept of invincibility that isn't really [there]."

This sense of security from homeowners is largely a by-product of effective corporate marketing. Like any viable market actor, private firefighting and foam application companies must generate demand for their product. And while most responsible businesses don't promise success, there is an expectation from clients that their services will in fact deliver tangible benefits—even in the face of firestorms. Efforts to rationalize private firefighting service transactions become difficult for homeowners because of the dichotomized nature of postfire outcomes: either a house survives a fire or it doesn't. There are very few ways that a paying customer will receive *incrementally* better outcomes than a foamless neighbor whose home burns to the ground. Therefore if the foam application process doesn't work properly clients will likely end up with a similar outcome as their neighbor. (Quite often partial burns end up requiring the same level of reconstruction as destroyed homes.) But despite these all-or-nothing outcomes private firefighting companies are still able to take advantage of wealthy homeowners' desire to attain increased security—even if that security is highly elusive in the context of a fast-moving, full-strength firestorm through tightly packed residential landscapes. These hard-sell marketing strategies in the face of significant product limitations in high-risk zones lead many high-ranking fire officials to question (and even outwardly disregard) compressed foam and gel application products. As one official noted while shaking her head, "The magic foam . . . it's kind of like, wow, snake oil."

To be clear there are substantive arguments supporting the situational use of foams and other gel products to fight fires. First, these products are capable of blanketing fuel surfaces more effectively than water can, thereby providing a durable layer to smother flames. Second, the foam or gel blanket is a useful mechanism for separating flames and ignition sources such as embers from the fuel surface. Although these products hold perhaps their greatest utility when extinguishing a flammable- or combustible-liquid fire (water is typically heavier than the hydrocarbon fuel it is being applied to, thus sinking below the surface), manufacturers argue that firefighting foams are also an effective defense strategy for home structures—they retain their fire suppression capabilities longer than water and are an effective method for buffering home surfaces from flames and flying brands. The problem of course is that real fire hazards and emergency home protection responses do not take place in a laboratory.

Gel application and compressed air foam systems are catered to homeowners in high fire risk areas either as a home kit or a full-service professional application process. Home kits such as those offered by the brands Cold Fire and FireIce make a hard pitch to potential customers and align do-it-yourself fire mitigation with forms of masculine heroism by labeling their product as "water on steroids." According to FireIce, the "Home Defense Unit" is a useful tool for homeowners seeking to protect their home from fire. "These days, water should not be the only tool used to fight an oncoming fire." The marketing material continues: "Application of FireIce will not only provide peace of mind, but the ultimate protection. Home fire extinguishers can help put out a small fire, but when large fires threaten your home, the FireIce will protect, suppress and extinguish serious fire risks."[6] Bold pronouncements of home security such as this are typical within the private firefighting marketplace. Armed with less than $1,000 and ample moxie, homeowners are led to believe they can defend their property in the face of an oncoming firestorm. This hard-sell strategy—"water on steroids" that will "protect, suppress and extinguish serious fire risks"—marks a technological fix to the problem of fire, a hyperbolized solution that conjures up images of "snake oil" for many members of the professional firefighting community.

Of further concern for many are the health and environmental effects of foam and gel exposure. Until recently many firefighting products contained or broke down into perfluorooctane sulfonates (PFOS), which are currently considered by the Environmental Protection Agency to be persistent, bioaccumulative, and toxic substances. In response to health concerns resulting from human contact with these chemicals, the United States along with several other nations have banned the production and importation of PFOS-based firefighting foams. As a result manufacturers are now developing a series of other products such as telomer-based foams that do not contain or break down into PFOS or other chemicals currently listed as persistent organic pollutants under the Stockholm Convention. Unfortunately the long-term impacts of perfluorochemicals in firefighting foams still persist in the environment. The extensive use of PFOS-based foams for training exercises at various military bases and airports across the United States has been linked to nearby groundwater contamination. Perhaps the most well-documented impacts occurred in Minnesota where perfluorocarbons were manufactured by 3M during the production of foam products.[7]

On Scene: Fire Mitigation That Increases Community Risk?

Of arguably greater concern to public emergency response agencies is the presence of private firefighting companies in neighborhood rescue settings. According to a fire official in Northern California, there are some instances when more hands on deck can be beneficial, particularly in an era where structures of mutual response are commonplace. However most of the time additional fire crews create new complications and problems that hinder the overall fire response. These are problems analogous to contemporary concerns over private drones flying in areas receiving active public air support. Private groups use drones to survey fires and locate at-risk homes. But these activities prevent response agencies from deploying water and fire suppressant airdrops due to concerns over midair collisions.

Concern about the presence of private firefighting businesses is particularly acute in cases when company employees have little or no prior training with local organizations or firsthand experience responding to fires in the area. "For most people ... once the fire happens you certainly don't want anybody in there like that, because it's a risk and you could have a potential situation that would impair your ability to take care of the incident," the official observes. She continues: "There's no standard of training for these folks, that is, you know, it's a private industry thing so ... it gets into a kind of dicey area. If you start letting a bunch of private contractors into an incident, it's not a good idea."

During the chaotic and oftentimes low-visibility environment that immediately precedes the main fire front, it can be difficult for fleeing residents and fire crews alike to differentiate private contractors from locally trained and seasoned firefighters. This presents a problem because many private outfits are unfamiliar with the local terrain and important evacuation routes. Moreover these crews arrive on the scene for the purpose of defending *individual homes* tied to insurance plans and *not* whole communities. As such they are not reliable sources of assistance for most community members. "They look the part," one fire official noted, before recounting a story from Southern California. "They were doing it down in LA . . . I mean, I remember years ago . . . all this dog and pony show, you know, with the home owners . . . and it was a mess down there because you didn't know who was who. And it gets confusing anyway because . . . you may have an engine from Orange County running around here and you're—as an agency—you're thinking well, okay, they belong here, they're part of something, you know, and they're not."

The contemporary neoliberal model of flexible governance and mutual aid only exacerbates this problem. Fire departments have become accustomed to the presence of outside agencies, diverse badges and crests, and unfamiliar faces. This diversification has fostered a level of trust between individuals, engines, and companies. In the hectic environment immediately preceding a fire, however, that trust can backfire. When certain responding units are not adequately trained in the local context and are charged with protecting only one or two homes in the area, the results can be catastrophic as residents are misled, roads are choked with private response vehicles, and general confusion over firefighting responsibilities reigns. The addition of sometimes oversized trucks within tightly packed emergency response environments can *increase rather than diminish* levels of risk during the evacuation process.

There are also explicit safety concerns for those actually employing fire suppressant materials. During a fast-moving firestorm there is very little time to apply foam or gel products, and it is impractical and dangerous for private company trucks to enter an active fire zone. These events are much different from isolated structure fires or airplane crash sites where there is ample room to maneuver, retreat, or modify firefighting tactics. In a residential community with winding streets, multiple ignition points, potentially high vegetation levels, and steep topography, there is very little (if any) room for error. As a result the most practical way to apply foam to homes is through prefire application. This means responding to early threats of fire and blanketing the home well before the fire hits. There are many challenges with this approach, most notably determining when to time retardant application so that it still contains active suppression capabilities when the fire strikes, yet is not applied so late that it places company personnel in danger. Timing is crucial here as once companies foam or gel a house the fire has to hit within a relatively short period of time; otherwise the substance loses its effectiveness. The length of its effectiveness depends on its drainage time, which is influenced by the amount of stabilizers present; the greater the amount of stabilizers, the longer fire resistance will last. According to Consumer Fire Products, most foams retain some level of utility for eight to sixteen hours, while gels typically retain their effectiveness for four to six hours.[8] This presents a very short window for private engines (most of whom will have to travel a long distance) to enter a threatened neighborhood or for homeowners to return home and apply fire-resistant products before quickly retreating.

Separate Service, Unequal Security, and the Atomization of Community Protection

Private firefighting services also raise ethical questions over who is best able to withstand the threat of fire. The head of Firefighters United for Safety, Ethics, and Ecology has suggested that while less affluent community members tend to rely on public services for home protection, this increase in privatization marks "a trend of wealthy providing their own protection." This line of criticism suggests that elite members of society can buy their way to superior protection through separate, expensive resource channels—a condition Naomi Klein labels "disaster apartheid," connoting a population clearly divided into secure and insecure segments.[9] Here some households have a marked advantage over their neighbors in home protection, leading to uneven fire exposure and recovery capacities. To be fair, while highly inequitable this is already a common condition under modern capitalism (i.e., neighborhoods already contain diverse levels of insurance coverage from one household to the next). However of even greater alarm is the fact that private firefighters entering a neighborhood will actually drive right past threatened homes and logical fire defense lines to find and protect the homes of their clients. This influx of private firefighting enterprises illustrates how sector privatization can lead to the atomization of entire communities. Fire mitigation (not just indemnification) becomes fragmented, stratified, and unequal at the neighborhood, street, and even individual homeowner level. The trend toward individualized property protection also runs contrary to traditional community-wide fire management practices. In an environment where fire spreads easily between homes and through entire neighborhoods, it is counterintuitive to privilege single residences for protection instead of defending the collective good. After all, every home is an asset and fuel source that if ignited could jeopardize even the most well-protected neighbor.[10]

Ultimately private companies—either paid for directly by individual homeowners or indirectly through premium insurance plan providers—will find themselves in an active fire evacuation zone (likely amid already chaotic conditions) without adequate knowledge of local social and environmental conditions, giving off an unclear authority status that may be confusing to others and charged with narrowly defined objectives that render mission ambivalence to other, perhaps more exposed homes. Whether one takes a critical or more sympathetic perspective of private firefighters and the commercialization of home products, one certainty remains: the private firefighting industry represents yet another group seeking to extract profits from high-risk residential landscapes across the West. New market opportunities continue to emerge within the private sector spurring new products, enterprises, and profits while also ushering in a new set of firefighting practices within suburban areas. The affluent nature of these communities—composed of homeowners with the capacity to pay more for additional firefighting services—only fuels this trend. The rise of private firefighting outfits and products provides further evidence of the concomitant production of risk and profits at the urban fringe. But perhaps more importantly the immense financial benefits, earning potentials, and ever-evolving risk reduction strategies held within these landscapes again raise serious questions about whether there is greater incentive to *encourage* rather than discourage further developments in fire-prone areas.

Conclusion

FROM EXCAVATING TO TREATING THE INCENDIARY

On June 28, 2013, a steady fire burned approximately thirty miles southwest of the city of Prescott in central Arizona. The fire spread gradually through terrain containing steep ridges, nearly flat valley bottoms, and a mixture of scattered rock piles, boulder fields, and chaparral brush. The rugged and in some places difficult to traverse landscape had not experienced fire in nearly forty-five years. Like much of the American West drought conditions and extremely dry vegetation, low-moisture content, high cured-grass loadings, and overgrown chaparral vegetation—in some cases over ten feet high—left the Yarnell Hill area highly susceptible to fire. A Drought Severity Index issued only one day before the fire described the region as having "extreme drought" conditions. By June 30, and coupled with intensified winds, the fire had grown faster and larger than was originally expected. Residents in the surrounding towns of Yarnell to the east and Peeples Valley to the northeast of the fire's origin were either immediately evacuated or placed on high evacuation watch. As the fire swept steadily through the parched landscape, incident command prepared a strategy to control the fire and eliminate or at least minimize its spread into the two adjacent residential areas.

The increasingly complex Yarnell Hill Fire required active mitigation on several fronts. As part of the multisite incident response a small group of firefighters known as the Granite Mountain Hotshots established a heel anchor along the southern edge of the fire. The team, part of the Prescott Fire Department, comprised twenty firefighters, all male, ranging from twenty-one to forty-three years of age. A Type 1 Interagency Hotshot Crew, the group of twenty were charged with actively fighting wildfires and engaging in efforts to reduce flammable vegetation growth in fire-prone areas, particularly in or nearby the wildland-urban interface (one member ended up serving as a lookout in a separate location). The team's anchor was designed to control the southern flank of the fire as it spread north and northeast from the team's position under prevailing winds. Generally speaking, a major component of fighting wildfires involves choking out the fire along its edges by stripping the landscape of its vegetation—either through manual vegetation removal or through careful and intentional burning techniques. The Granite Mountain team had performed well and the southern anchor position was holding as planned.

As the day wore on, the Granite Mountain crew watched the fire closely from their position atop a ridge near the southern edge of the fire. Although the fire had generally burned away from their position, a nearby thunderstorm outbreak was threatening to produce an outflow boundary that would bring with it steady gusts of wind spreading in a more southerly direction. Sure enough, within a few hours the outflow generated intense winds, which brought significant behavioral changes to the fire. This included a ninety-degree change in direction and a rapid increase in the speed of the fire's spread. Before long the head of the fire had shifted. It was now moving back in the direction of the Granite Mountain crew.

Noting this change in fire movement, the group left the black (connoting an already burned area) and descended the ridge through unburned slopes to their south. They headed toward a designated safe area at the Boulder Springs Ranch located at the base of the rocky hill area. As they dropped in elevation, the fire fell from view for several minutes as it became obscured behind the descent route topography. Once they emerged from behind a series of rocky knolls, the fire front, which was now moving at a rate of ten to twelve miles an hour, had gained on them significantly and had effectively cut off their route to the safety zone. With the now ferocious blaze bearing down on them and faced without any other good contingency options, the crew was forced to hastily use their emergency deployment fire shelters along a flat area at the head of a box canyon. Members of the crew even tried to quickly burn the area around them—an attempt to remove nearby vegetation and protect themselves from direct fire contact. But the flames advanced too quickly, overtaking the emergency shelter deployment just as it was being established and hitting the firefighters with temperatures above two thousand degrees Fahrenheit. All nineteen firefighters that were part of the emergency deployment perished in the fire. They were only six hundred yards from the safety zone.[1]

The Yarnell Hill Fire was the most deadly fire in terms of firefighters lost to a wildfire since 1933 and the deadliest wildfire of any kind since the 1991 Tunnel Fire in the Oakland Hills.[2] In typical fashion several debates ensued immediately after the fire. These included questions over whether greater federal funding would have helped control the fire (particularly in securing greater aerial tanker assistance) and whether the areas being protected were actually defensible (despite Yavapai County's Community Wildfire Protection Plan) given the dry, rugged terrain and poor access road infrastructure. But what these debates largely failed to address were larger questions over what precipitates (still today) the spread of human structures into fire-prone areas and perhaps more crucially what role and *risks* firefighters should be expected to take when confronting such fires.

The towns of Yarnell and Peeples Valley, which were directly impacted by the fire, have less than one thousand residents. Since their original founding by gold prospectors in the late 1800s these two settlements— approximately thirty miles southwest of Prescott—have avoided significant population growth. The areas threatened by the Yarnell Hill Fire are therefore not your typical suburban sprawl characterized by newly established subdivisions at the fringe of more established urban development. And yet the case of the Yarnell Hill Fire offers an important lesson about wildfires in the West, a lesson that holds direct relevance to contentions put forth in *Flame and Fortune*. The plight of the Granite Mountain Hotshots suggests that arguments in support of exposing and reversing the causes of extensive urban encroachment into undeveloped fire-prone areas are not frivolous, petty, or unwarranted. They are not arguments

188   CONCLUSION

against abstract risks and costs. Rather these are concerns motivated by real, tangible vulnerabilities associated with the active defense of communities threatened by wildfire. The nineteen Granite Mountain Hotshots—sons, brothers, and fathers—who perished on June 30, 2013, are a prime example of these very real and embodied risks.

The presence of residential settlements and our inclination to defend them is why we send firefighters so aggressively into harm's way. Where homes go, risks will follow—and not just for the homeowners. While our culture has come to expect and even celebrate such heroism, these are risky activities that we should actively seek to minimize. Why should young men and women across the West risk their lives to defend homes that are constructed to serve the interests of profit-seeking developers and amenity-seeking residents? This question rings especially true for areas built in recent decades when knowledge about increased fire risk is well established—that is, in areas where residents come to assume fire risks yet often shirk any sense of seriousness (see Figure 25) about these hazards due to love of landscape and, for most, the availability of considerable resources for offsetting risk. (Of course not all residents should be viewed as privileged, negligent, and sufficiently indemnified. There exist tangible vulnerabilities, even amid affluence, which planners and policy-makers will need to grapple with.) Even the insurance companies are subsidized in a sense from the full cost of paying (and thus charging occupants) for fire mitigation services as state and federal governments cover much of the firefighting burden. What responsibilities do *nonresidents* have to risk their lives protecting these residential areas? What responsibilities do developers and city and regional planners have to slow and even cease the construction of residential developments and thereby limit opportunities for still further fire response tragedies? These are no doubt complex questions without easy answers.

We are facing a crucial moment in our region's planning history. The 2013 Yarnell Hill Fire, like the 1991 Tunnel Fire, highlights real risks and true tragedies that exist when fighting fires at the wildland-urban interface. *Flame and Fortune* has endeavored to show various mechanisms and entities that extract wealth from these landscapes—processes that in turn facilitate the development of WUI environments and risky landscapes around the West. Understanding the Incendiary and the affluence-vulnerability

CONCLUSION 189

Figure 25. A satirical depiction of the modern West displaying the relationship between vulnerability and affluence for homeowners armed with considerable risk-offsetting resources. Though the image is sardonic in tone, real-life response conditions remain serious as firefighters and others are placed in high-risk circumstances when defending properties from the threat of fire. (Credit: Tracy Stuckey)

interface more generally presents an opportunity to first imagine and then implement a strategy to combat the root causes of development at the urban periphery.

It is time we combine our interest in landscape outcomes with a strong commitment to socioecological processes. No longer should civil society and politicians alike allow profits to be extracted from landscapes at the metropolitan fringe as if there were no consequences. No longer should cities and developers be able to so easily conform to fiscal pressures and incentives for continued urban expansion as if real trauma and loss will not follow. No longer can we merely treat the symptom of the problem by tinkering with small fires within the WUI. Instead we need to address the root causes of the firestorm itself by treating western WUI areas like a patient—the Incendiary—through a comprehensive assessment of landscape backgrounds, histories, underlying drivers, internal governing mechanisms, core characteristics, and externally influential forces. Although no single template will solve the problem, a few strategies outlined below will serve to initiate a conversation about how to shift our attention from grappling with WUI *symptoms* to tackling AVI *processes*.

· · · · ·

Three broad approaches to contend with costly and deadly suburban and WUI fires are detailed below. These approaches are by no means exhaustive and are presented here as a way to categorize the many policy and planning tools at our disposal. Ultimately political and fiscal constraints will necessitate a combination of these and other approaches.

REFINING AND EXPANDING COMMUNITY ADAPTATION

The first approach is quite straightforward and entails continuing our current activities and approaches only with more extensive implementation. This largely "business as usual" approach relies heavily on fire adaptation, which is premised on community preparedness training, education, and the enforcement of city and county building codes, zoning ordinances, restrictive covenants (such as those tied to homeowner associations), and vegetation management regulations. It will entail combining public resources—funded by state, county, and local-level revenue-generating mechanisms—with increased levels of private financing in the form of household fees and for-profit fire mitigation service providers. A central component will be prefire preparedness and not simply reactionary decisions that arise after deadly fire events. Within this vein regionally coordinated efforts that align administrative capacities and leverage high-impact potentials from funding agencies have proven successful.

The City of Oakland, and the Tunnel Fire area more generally, is an example of an area containing both far-reaching and inventive fire preparedness measures. These range from property compliance regulations to coordinated and multiple-municipality governance regimes. The trouble with Oakland's compliance and inspection approach is that it arrived several years too late as a response to the Tunnel Fire. Efforts under the "Refining and expanding community adaptation" approach should take a more proactive approach and aim to prevent disasters from arising (or at least minimize their likelihood) in the first place. Clearly if new suburban developments are to be constructed they should be forced to follow strict suburban zoning ordinances, covenants, and building codes.

The property compliance category of fire preparedness in very high risk (VHR) fire areas of the Oakland Hills for example involves sending all area residents pamphlets outlining strategies for implementing adequate defensible space around homes (see Table 2). These guidelines are followed by approximately 26,000 public and private inspections annually, many of which result in fines, fees, and other abatement costs. Private properties are divided into three zones. Zone 1 comprises land and structural features within zero to five feet from buildings; these California codes are designed to reduce the chance of windblown embers igniting materials near homes. Mandated practices in Zone 2, between five and thirty feet from buildings, are designed to create a landscape that will not readily transmit fire to the home. Zone 3 modifications are located thirty to one hundred feet from buildings and aim to reduce the energy and spread of wildfires. These codes and annual inspections are complemented by periodic "red flag days." During times of high wildfire risk (characterized by low moisture content in vegetation and the threat of Diablo winds) firehouses in the area raise a red flag. Accordingly residents are reminded in pamphlets distributed at the beginning of each summer to (1) park cars in driveways to increase first responder access, (2) update and review household disaster and evacuation plans, (3) coordinate with neighborhood-based groups established within the "Communities of Oakland Respond to Emergencies" (CORE) program, and (4) clear dead vegetation around structures and in gutters. Among other restrictions

Table 2 Sample elements of the "Refining and expanding community adaptation" approach to confront the Incendiary. An example of city property fire prevention plan and associated codes (City of Oakland). CFC = California Fire Code; CBC = California Building Code. (Source: 2015 Annual Inspection Notice, Fire Prevention and Support Services Bureau, Oakland Fire Department)

Zone	Zone Description	Zone-Specific Codes
1	**0–5 feet from building:** Reduce the chance of windblown embers igniting materials near home, exposing it to flame	Maintain and clear shrubs and brush of dead vegetation (CFC 4907.1). Choose Class A–rated roof covering and remove debris from roof on a regular basis. Trim tree limbs within 10 feet of roof (CFC 4910.2.2.4). Clear rain gutters of dead leaves and debris (CFC 4907.1.5). Remove ivy and any vines climbing on house or trees, which act as "fire ladder" (CFC 4907.1.4). Spark arrestors required on all fireplaces (CBC 3102.3.8). Stored firewood at least 20 feet from structure (CFC 4907.1.1). Cover attic, foundation, and other vents with 1/8-inch metal mesh screens. House number must be clearly visible from street, at least 4 inches in height, and in a contrasting color (CFC 505.1). Maintain 6-inch clearance between ground and start of home siding.
2	**5–30 feet from building:** Create a landscape that will not readily transmit fire to home	Cut grass to 6 inches or less (CFC 4907.1). When clearing vegetation, use care when operating electric- and gas-powered equipment such as lawnmowers; one small spark can start a fire; a string trimmer is safer. Large trees do not have to be removed as long as plants beneath maintain adequate spacing and eliminate vertical "fire ladder" (CFC 4907.1.4). Remove tree limbs within 6 feet from the ground (CFC 4910.2.2.4). Wood chips must be less than 6 inches deep; no piles (CFC 4907.1).

3	**30–100 feet from building:** Reduce the energy and speed of the wildfire	Maintain plants and shrubs under trees to eliminate a vertical "fire ladder." Maintain a defensible space, clearing dead brush and debris. Be aware of the vertical and horizontal slopes on properties and use proper spacing to prevent "fire ladders." When clearing vegetation, use care when operating gas-powered equipment; one small spark may start a fire.

residents are also not allowed to use gardening equipment that could set off sparks, have barbeques, or establish campfires.

Along with land- and property-based management, current fire mitigation involves close coordination between adjacent cities, municipal districts, and major public landholders. In the Tunnel Fire area this organization is known as the Hills Emergency Forum (HEF) and contains member agencies from the cities of Oakland, Berkeley, El Cerrito, and Orinda/Moraga as well as Cal Fire, East Bay Regional Park District, East Bay Municipal Utility District, University of California Berkeley, and Lawrence Berkeley National Laboratory. Close consultation also occurs with fire chiefs from the nearby cities of Albany, Fremont, Hayward, Piedmont, and Richmond. Together these member and consultative agencies

> coordinate the collection, assessment and sharing of information on the East Bay Hills fire hazards and . . . provide a forum for building interagency consensus on the development of fire safety standards and codes, incident response and management protocols, public education programs, multijurisdictional training, and fuel reduction strategies.[3]

According to several members on the HEF Staff Liaison Committee, one of the main benefits of this mutual action group is the ability to coordinate proposals for state grants. This includes working together to generate a unified set of integrated and mutually enhancing mitigation objectives and implementation timelines. Though not without its challenges this brand of interjurisdictional fire management has proven successful, as one member put it, simply "because it's productive for the staff to meet!" Shared governance is a crucial component of fire management in the Oakland and East Bay hills that has expanded and harmonized regional community hazard preparedness. Other regions lacking such coordination would do well to pursue or bolster such collaborative efforts.

While these current efforts have seen considerable success—and the individuals organizing and implementing fire mitigation deserve praise for their hard work with limited resources—there are noteworthy drawbacks that come with this approach. We have seen that city revenues to support these kinds of local fire mitigation activities have historically been generated paradoxically by building more homes in fire-prone landscapes! Of course this just adds to the scope of the problem. And increased reliance on private sector support has led to the implementation of firefighting activities that present challenges for public firefighting outfits (and undesirable and inequitable outcomes for certain homeowners). Furthermore without adequate investment in municipal infrastructure important modifications such as water piping upgrades and the placement of power lines underground will be difficult to achieve. And without these fundamental upgrades—particularly in older, less affluent neighborhoods—many of the benefits normally derived from community education, building code enforcement, and other fire preparedness measures will be limited. If increased adaptation is the goal then additional city financing of road, water, electric, and other base resources will be necessary to actuate effective community fire preparedness across the region.

Perhaps most importantly, deepening fire adaptation means eliminating the creation of *new* fire risks. There is nothing rationally adaptive about intentionally producing more vulnerabilities. A concerted effort should therefore be made to prevent further home developments in already high-risk areas. Fire adaptation is certainly important in areas with already existing residential communities. Indeed it is argued in this approach that similar measures should be implemented earlier, more frequently, and with greater enforcement capacity. However, as the sections below detail, there is also a need to move beyond fire prevention and develop strategies to slow the Incendiary and stop the production of new homes and new risks in currently undeveloped areas. Merely adapting to an arsonist is not an effective long-term strategy for reducing risks and unnecessary costs.

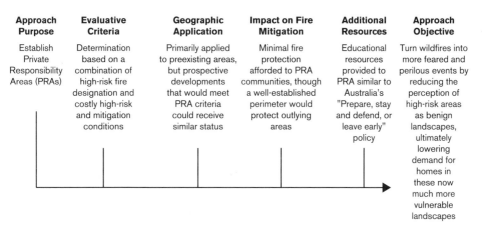

Figure 26. Core elements of the "Fighting fire with fire" approach to combat the Incendiary.

FIGHTING FIRE WITH FIRE

A second method to reduce new housing developments involves increasing risks associated with living in high fire activity areas. Such an approach will entail establishing private responsibility areas (PRAs) or some other similar land designation connoting areas with very high risk (VHR) where firefighting crews are put in grave danger and where the financial costs of active fire response efforts are particularly high. In these areas minimal fire assistance would be provided to residents (see Figure 26 for outline). Only regions of highest fire risk would be designated in this way and training would still be provided to educate these communities on key fire preparedness strategies. Given the reduction in first responder assistance, these education programs would more closely resemble Australia's "Prepare, stay and defend, or leave early" policy, which trains residents on how to deal with fire danger in the absence of a large-scale public fire response. Fire perimeters would be established to enable a more comprehensive response to wildfires that burn beyond PRAs.

This may appear a draconian response and one that resides on the opposite end of the spectrum from the aforementioned adaptation approach. Social tractability and political feasibility are certainly not its strong suit. But there is a logic here suggesting (as Mike Davis does in

"The Case for Letting Malibu Burn") that the benefits of firefighting are simply not worth the financial costs and social risks. Emergency response handbooks typically advise against creating additional victims during rescue and protection operations. The case of the Yarnell Hill and Tunnel fires illustrate that firefighting is not a risk- or consequence-free endeavor and that second (or significantly more) victims are frequently generated by the emergency response itself.

There are corollary fire behavior benefits that could arise under the "Fighting fire with fire" approach. We know that one of the major drivers of current-day, high-intensity fires is our nation's long history of fire suppression. Regular, naturally burning fires will tend to promote fire regimes comprising low- to moderate-level blazes. Fire exclusion, on the other hand, has created landscapes across the West that are more susceptible to large and more intense fires.[4] Historical fire suppression was in many cases justified as a way to preserve the integrity of valuable forest stocks in our national forests and other public and private lands. But our desire to prevent and quickly extinguish fires was also driven by efforts to protect another suite of prized assets: city infrastructure, residential communities, and individual homes.[5] If we more readily let fires run their course and closely mimic natural fire activity, then the propensity for large and intense fires that eviscerate hundreds of thousands of acres at a time should be decreased.

Perhaps most importantly this approach offers something different, even radical, by offering fire as an antidote to fire. As wildfires become more dangerous, potent, and perilous in the public eye, involved parties (including potential home buyers) may wish to avoid them altogether. There is no question that the urbanizing West is not in the business of generating ghost towns. Already developed areas at the city periphery are here to stay, at least for the medium term and until some other resource limitation (such as continued water shortages) forces widespread city compaction. The objective of this approach is rather more performative in nature. It is intended to reveal adversity and demonstrate the immense hardships, anxieties, and costs associated with living in these areas. It is therefore meant to reduce demand for construction in areas that are rendered outside the purview of public assistance. Other benefits would include a sharp reduction in fire mitigation costs and firefighter risks.

If residents assume elevated fire risks by choosing to live in locations already prone to fire (and as this book suggests, increase fire susceptibility simply by virtue of their presence), and if they are willing to pay for this risk through the purchase of comprehensive indemnification, then it is not unreasonable to view massive public expenditures as both enabling and unnecessary. This is the prevailing logic of the "Fighting fire with fire" approach. By excluding these communities from conventional public fire mitigation services, wildfires may become more feared events. Large fires could become a threat more households actively seek to avoid. Over time a lack of demand for homes in these now much more isolated and vulnerable landscapes may follow, particularly as stories of hardship for those without premium insurance begin to propagate through neighborhoods and news outlets. In short, if the goal is to stop further developments then current landscapes cannot appear to the public as benign landscapes full of charm and bereft of substantive risks. This approach suggests that thwarting the spread of WUI areas further away from the metropolitan core will only occur if fire is seen in a more formidable light.

Of course drawbacks exist with such an approach. For example the establishment of PRAs could disproportionately impact poorer and minority households lacking sufficient insurance coverage or rainy-day fund resources. Many researchers such as Tim Collins have described how marginal groups (such as those in trailer parks at the urban fringe) can be negatively impacted by planning efforts that cater to wealthier, amenity-seeking residents with sufficient support and indemnification plans in place. Moreover as the case of the Tunnel Fire demonstrates, households are heterogeneous in their histories, financial status, and capacity to recover from wildfires. Not all members of a community share the same level of risk sensitivity. The presence of marginal communities and nature of variegated household risk profiles may very well require the implementation of special exclusion areas for neighborhoods falling beneath preestablished economic thresholds, such as average annual household income or average property value across a geographical area. Adequately serving individual households with the highest net vulnerability (i.e., those with elevated exposure and reduced capacity to respond to fire—along financial, social, psychological, and geographical lines) would be extremely difficult yet necessary if this approach were to be pursued in earnest.

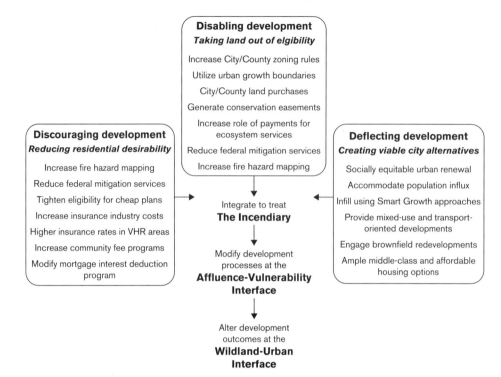

Figure 27. Core elements of the "Reducing demand for development" approach to combat the Incendiary. Elements of this strategy will likely be enhanced if implemented alongside aspects of the "Fighting fire with fire" and "Refining and expanding community adaptation" approaches.

REDUCING DEMAND FOR DEVELOPMENT

A third strategy for combating the Incendiary involves reducing overall demand for new housing subdivisions at the urban periphery through diverse land acquisition, zoning, and home finance policies while also redoubling sustainable urban in-fill development practices (see Figure 27). This approach provides somewhat of a middle ground strategy that will naturally incorporate elements of both the "Refining and expanding community adaptation" and "Fighting fire with fire" frameworks. The goal of the "Reducing demand for development" approach is to simply make it more difficult to implement further suburban development in VHR areas. There are three aspects to this approach: (1) discouraging development,

(2) disabling development, and (3) deflecting development. These elements are closely linked and would be implemented most naturally within a single integrated approach.

Discouraging Development

Discouraging development in high-risk fire areas involves making it much less practical and logical from a cost-benefit perspective to convert nondeveloped land into developed residential tracts. Efforts to disincentivize development in high fire risk areas will include a combination of options that include fire mapping and higher insurance premiums. Rather than take areas out of eligibility altogether (see Disabling Development below), this approach simply makes it less desirable and potentially more expensive to engage in development. When the cost burden of living in these areas goes up then demand should go down. And when the appeal for potential home buyers is reduced demand for home construction should diminish as well.

MAPPING FIRE-RISK SEVERITY

An integrated "reducing demand for development" approach begins by identifying and delineating high-risk fire areas and subsequently implementing land use management approaches that impede future developments in these already fire-prone areas. Cal Fire's Fire Hazard Severity Zone (FHSZ) mapping program, which has been utilized in various institutional capacities across the state, is exemplary of this approach. In general the availability of more detailed and affordable mapping technologies in government and private sector settings makes this an achievable strategy element, especially in comparison to past decades when geospatial and remote sensing limitations hindered detailed fire risk mapping. Hazard zones are determined through the use of models that predict the likelihood and severity of fires and are based on topographic, climatic, vegetation, and ember-production characteristics and on historical fire records. Once designations are created and mapped, ordinances can follow. These include regulations requiring that structures in high hazard severity zones are built using fire-resistant features identified in California (or other strict) building codes. These designations can also serve local communities as an effective and ongoing education and public awareness tool.

The existence of fire hazard designations may also assist efforts to prevent developments altogether, although fire maps alone will not automatically trigger new antisprawl policies. Instead comprehensive hazard mapping must serve as the basis for a new and aggressive suite of fiscal, planning, zoning, and fire mitigation policies at local, state, and federal levels. In Napa County, California, for example new buildings and subdivisions can be legally rejected due to their location in high fire risk areas and as a result of proposed placement in landscapes with steep topography, dense vegetation, and inadequate infrastructure for firefighting.[6]

INCREASING THE FINANCIAL BURDEN FOR RESIDENTS:
INSURANCE AND OTHER HOUSEHOLD COSTS

Such risk labels will do much more than merely connote risk geographically. First, they can cause insurance companies to cancel plans for residents in VHR designated areas. (While conducting research for this project, I interviewed several people who had their insurance plans canceled because they live in high-risk areas. This includes a compliant and fastidious Northern California fire marshal who received a call one day indicating she had been dropped by her carrier!) The mere threat of lost coverage could deter certain individuals from buying homes in high-risk areas.

Second, those who remain with insurance companies in designated high-risk areas are likely to see higher coverage rates. Significantly raising rates will act as a deterrent to new home construction but will only be achieved if insurance companies have a compelling reason to charge higher premiums. One way for this plan to gain traction will be to remove much of the protection costs provided by federal taxpayers. This would include cutting back plans like California governor Jerry Brown's 2016 budget earmark for a $719 million one-time drought package and extra $215 million to the state's emergency fund to fight wildfires in the state. Currently the financial risks associated with insuring homes in the WUI is far less than it would be without the Bureau of Land Management (BLM), FEMA, and U.S. Forest Service paying a large portion of the costs to protect residential areas. With less assistance from these groups (by for example combining "Reducing demand for development" with a version of the "Fighting fire with fire" approach and its use of PRA designations—see above) the cost burden of devastating fires will be higher for insurance companies, which will now be responsible for compensating a higher level of annual disaster recovery costs back to fire victims. A need for elevated household insurance premiums should follow. (A modified and less strict version of this plan would only charge higher premiums to those households that fail to maintain adequate defensible space or implement other fire-safe regulation requirements.)

Third, linking land use decisions to recognized risk designations could provide the basis for eliminating mortgage interest deductions. This would involve legislating that new homes built in VHR areas of the WUI do not qualify for such financial incentives.[7] The mortgage interest deduction has the effect of encouraging particularly well-off home buyers to purchase bigger, more expensive homes oftentimes on larger lots in order to increase the total transaction cost, mortgage size, and future interest savings. Eliminating the mortgage interest deduction for new residences built in designated fire-prone areas could potentially reduce the size of homes while also lowering residential demand in these high-risk zones. In turn this could reduce both future home developments and fire suppression costs. Given the mortgage interest deduction's central position within America's entrenched homeownership society, this could prove a difficult piece of legislation to modify. (If primary residences wind up being a difficult target, then second homes may be a more tractable policy goal.) Despite potential challenges all options should be left on the table as we gain better recognition of which homes and communities are dependent on U.S. taxpayers for routine fire mitigation activities.

Like the aforementioned federal fire mitigation programs offered by the BLM, FEMA, and the Forest Service several state and local programs also seek to reduce insurance costs for homeowners through subsidies and other cost-share programs. Unfortunately many of these approaches have been clumsily implemented resulting in outcomes that undercut efforts to slow the Incendiary. For example California's FAIR Plan offers cheap insurance options for state residents. While the program's overall objective is laudable—providing inexpensive insurance choices for those who can't afford the going market rate or for those recently dropped from a policy—it can in some cases also make it quite inexpensive to live in high-risk areas. A more nuanced FAIR Plan would carry more stipulations, including providing discounted rates only in cases of emergency, high financial insecurity, and in areas with preexisting homes, thus disincentivizing demand for new developments. Along with reducing government support mechanisms in WUI areas, household-level fees can be implemented that provide firefighting resources paid only by the households benefiting from resulting services (see the SRA and WPAD fee debates in Chapter 7).

Each strategy to increase overall insurance costs for high-risk area residents—reduced federal assistance, modified state plans serving only the most disadvantaged insurance holders, and the implementation of household fees—will also insulate other citizens (that use similar providers) from carrying the cost of subsidizing residents in fire-prone areas. This is the nature of a multiscaled and integrated anti-Incendiary plan: it will require creative and coordinated efforts from public, private, and non-profit sectors at multiple scales.

By ushering in potentially higher home insurance costs and reduced public fire protection, comprehensive hazard designations may also have an impact on short-term home sales transactions. In the State of California, for example, disclosures are required for all home sellers indicating very high fire risk designations to potential home buyers. Evidence suggests that living in such areas could make sales transactions for residents in these areas more expensive in the immediate term.[8] A 2006 study showed that since the 1998 disclosure requirement went into effect the average price of homes in very high risk areas was 3 percent higher than for comparable homes outside the area.[9] As shown in Chapter 9, the real estate market in the Oakland Hills is as lucrative as ever. Consequently the effect of such designations would be to attract only those residents who are willing to "go it alone," so to speak, and pay more for elevated insurance costs and private firefighting services. (This is different from arguments put forth earlier opposing private firefighter involvement in mixed-class/coverage neighborhoods and in the presence of public firefighter activity.) Such an approach suggests that many VHR areas could over time become suited only to those with the financial resources to fund their own protection. Of course partitioning society into "haves" and "have-nots" is not a desirable outcome in most cases. However in situations where large homes already exist an argument can be made in support of at least minimizing the responsibility of average homeowners to subsidize elite members of society.

Disabling Development

Disabling development entails implementing various strategies that will make it difficult if not impossible to convert open space into housing developments and other suburban land features. Essentially this will involve removing developable lands from eligibility and will include designating more areas as parks, green belts, and conservation easements. New zoning rules will also remove certain areas from suitability. Most of these efforts would conform with contemporary "smart growth" approaches which are premised on the construction of compact cities with easy and close movement between work, school, commerce, and home. This approach thus advocates for less outward sprawl through the use of urban growth boundaries that discourage new suburban developments in VHR zones.

CONSERVATION EASEMENTS, ZONING ORDINANCES, AND PAYMENTS OF ECOSYSTEM SERVICES

On their own, techniques of discouragement will not be enough to curb the Incendiary. Taking whole tracts of land out of development eligibility will be required as well. One such disabling strategy is to convert areas into conservation easements. This is a particularly effective approach in areas currently without housing subdivisions and where the cost of providing future fire suppression services is projected to be high; that is, in areas that are remote or where current resources would require the purchase or rental of new and additional costly mitigation assets. The case of Plum Creek, Montana, provides a good example. Here, in the face of recent land sales for residential developments, Montana legislators agreed on a bond for $21 million to purchase a block of threatened land in Missoula County. The program, labeled the Working Forest Project, determined that developed lands could increase wildfire suppression costs by up to $73.7 million if produced at a density of at least one household per

160 acres. The $21 million Working Forest Project approach turned out to be a much more cost-efficient approach for managing the landscape ($21 million versus $73.7 million) even when factoring in potential revenues derived from new property taxes.[10] These long-term financial benefits may persuade other governments to purchase areas for conservation and would provide effective talking points for nongovernmental groups like the Nature Conservancy, the Trust for Public Lands, and other groups active in the increasingly popular easement business. Counties such as Boulder (CO) have also seen success buying open space and conducting land swaps with federal agencies in order to protect land from further development.

Zoning ordinances present another disabling approach. Here local governments establish zoning rules that influence where future homes can be built. If structured properly, resulting land designations can effectively redirect construction activities away from fire-prone areas. Areas identified as agricultural, parkland, or some other development-exempted zoning status would be off-limits to residential construction. This type of zoning regulation could be implemented as part of a broader urban growth boundary initiative. (But as in the previous section, involvement by local agencies will only occur if federal governments cease carrying such a large percentage of the fire mitigation burden.) Currently most local governments encourage participation in fire-adapted community activities, such as maintaining defensible space around homes and ensuring the enforcement and establishment of new building codes. However most of these ordinances, covenants, and regulations do not actively take measures to prevent further developments on fire-prone lands. (There are exceptions of course. Counties such as Napa [CA] and Skagit [WA] are examples of areas that are aggressively implementing strategies to minimize new developments through creative rezoning measures.)

For property owners deemed outside the area of allowable outward urban development, a potential negative impact may be a reduction in property values (or at least future value growth). Indeed this is a critique of urban growth boundaries more generally, as certain areas redlined as "nondevelopable" hold in a sense a much lower rent-gap value. (This means the current value of a property, particularly rentable property, is somewhat comparable to its potential highest-achievable property and rental value. A large rent-gap situation, on the other hand, occurs when a significant difference between current and future values exists, assuming the addition of sufficient investment.) One way to make growth boundary zoning approaches more acceptable to private landowners and city and county agencies linked to nondevelopable areas will be to clearly articulate the potential benefits of payments for ecosystem services (PES). The emergence of PES provides a financial incentive to convert developable lands into new "hands-off" land status, such as a conservation easement. Through this change in land status and management all parties contribute to an "avoided" negative outcome (i.e., by removing the possibility of deleterious future development activities on nearby ecosystems). The establishment of conservation easements will become a more desirable land use choice for landowners once they realize they can receive payments in return for preserving important ecosystem functions, such as increased water purification, carbon sequestration, species/ecosystem protection, and viewshed preservation.

Deflecting Development

Deflecting development involves reorienting land conversion activities away from suburbia and turning the course of housing construction back toward city centers through socially equitable urban renewal. Realistically, given high population growth rates in western cities, a reduction in further suburban developments will only occur if a viable alternative is presented. Urban renewal, city in-fill practices, and a strong commitment to reinvesting in metropolitan core areas (and away from periphery landscapes as has been customary over the past half-plus century) will be necessary to create vibrant and desirable places to live. These efforts will continue to present new and alternative residential options for current and future city occupants. It will be absolutely crucial however that urban renewal approaches proceed in a manner that produces socially equitable outcomes catering to a broad swath of class, race, and gender categories. This will ensure that planning efforts do not disproportionately place marginalized populations in marginal or dangerous landscapes.

PROVISION OF HOUSING ALTERNATIVES THROUGH SOCIALLY EQUITABLE CITY IN-FILL AND URBAN RENEWAL

The aforementioned approaches—designed to discourage and disable new developments—will only gain traction if other viable urban housing options are made available. Like most other areas of the world the western United States is becoming an increasingly urban region with metropolitan areas experiencing a 63.6 percent increase in population between 1980 and 2006.[11] It appears this trend is only intensifying over time. If relatively fewer residents move to the urban periphery, where will they go? Given this great urban migration, it will be critical for alternative housing markets to be provided that deflect development away from the suburbs and turn the flow of housing construction projects back toward city centers.

Providing space for this influx of city residents will begin by reinvesting in core urban residential areas around the West. This will entail the implementation of neighborhood renewal and city in-fill practices such as brownfield redevelopments and mixed-use housing construction in currently undeveloped, former industrial, or other blighted areas within central city areas. Increasing public and alternative transportation options such as light rail and expanded bus routes as well as greater car and bike share programs will be necessary to accommodate more residents in these now more densely populated urban environments. Transportation-oriented developments will provide a useful way of integrating mixed-use housing approaches with public transit by strategically positioning new residential construction along transportation corridors and next to transit stations.

Examples from the around the United States however show that massive investments in urban renewal (also referred to as gentrification) can wind up alienating preexisting, poorer communities in favor of more affluent residents who are able to accommodate higher real estate values and rental costs. Under this scenario disadvantaged residents are pushed into outlying areas, as are the cultural heritages, practices, and artifacts they created over many decades. Moreover in the West these outlying areas at the suburban fringe might include districts with smaller homes and containing unstable environments (with for example high exposure to fire activity). There are many domestic and international examples of areas that once deemed marginal and dangerous become sites for cheap land occupancy. These include flood- and landslide-prone and contaminated areas that are often converted into slums or other substandard living options. These areas attract residents who not only become disproportionately exposed to hazards but also do not have the economic means of recovering from inevitable catastrophic events.

This is clearly an unethical and unacceptable outcome. A concerted effort will be needed to ensure that urban fringe areas are not rezoned for such low-cost housing. And more importantly urban renewal efforts in the city core will need to be achieved through socially equitable approaches that incentivize city in-fill *while also* including provisions (such as inclusionary zoning programs) that mandate a certain percentage of low- and middle-class living options and preserve the cultural legacy and social dynamics of preexisting city neighborhoods.

FLAME AND FORTUNE: A LOOK BACK AND A WAY FORWARD

Ultimately governments and citizens of the American West as well as other fire-prone areas of the United States and abroad must do more than treat symptoms of the Incendiary. Fire adaptation is certainly a worthwhile cause that should be encouraged across the region. However as costs mount and lives are lost, we must do more to undercut the drivers of rampant sprawl into already fire-prone areas. A 2002 study by the U.S. Fire Administration estimates that in the American West "38 percent of new home construction is adjacent to or intermixed within the WUI."[12] More crucially the possibility for still greater expansion is a distinct possibility given that only 14 percent of western forested private land adjacent to public land is currently developed for residential use. This leaves tremendous potential for future development on the remaining 86 percent. As populations in the West rise and as cities expand, it will be illogical to continue intentionally producing the conditions we readily fear, the conditions we seek to minimize, and the conditions we already spend millions of dollars adapting to each year. Tackling the affluence-vulnerability interface must become a priority and such efforts should begin by exploring ways to disable, discourage, and deflect new developments at the urban periphery. This will require a three-pronged approach that includes managing preexisting residential areas in high-risk zones differently in order to discourage future demand, reallocating currently eligible lands for other purposes, and generating alternative housing options for all within the city core.

The Tunnel Fire, which has served as a central thread throughout *Flame and Fortune*, illuminates various ways land developments at the urban periphery (even a century ago) are driven by rapacious financial opportunism and profit-seeking motives. A close analysis of the Tunnel Fire area (along with examples from other areas in the American West) demonstrates the systemic and entrenched drivers producing and maintaining these lucrative, vulnerable, and costly landscapes. It has been said that "the past is prologue." Indeed the structural root causes producing high-risk fire conditions in the Oakland Hills continue to influence landscapes and drive developments across the region. The Tunnel Fire and the forces behind its making remain as significant as ever, even twenty-five years after the event.

Given the diverse ways political debates, community disagreements, and mainstream and scientific discourses on fire conceal these troubling developments, *Flame and Fortune* has sought to excavate and highlight key dimensions of the affluence-vulnerability interface and its symptom: contemporary manifestations of the wildland-urban interface. It has also been said, "One should only offer solutions if one first understands the problem." Accordingly, only after diligently examining the Incendiary and being armed with a keen understanding of the persistent, complex, and underlying machinations of the AVI can we begin to conceptualize and articulate potential solutions within the WUI. Understanding these drivers and incentives, with the use of a detailed case study of the Tunnel Fire, has been one of the book's principal aims. The path forward will certainly not be easy to conceive or attain, but this does not mean we should treat the road ahead as inconceivable or unattainable.

Notes

CHAPTER 1: THE 1991 TUNNEL FIRE

1. Federal Emergency Management Agency. 1992. The East Bay Hills Fire Oakland-Berkeley, California. U.S. Fire Administration/Technical Report Series USFA-TR-060, Washington, DC: FEMA.
2. According to the National Fire Protection Association (2014) the most costly modern U.S. fires overall (in adjusted 2013 dollars) are (1) the September 11, 2001, World Trade Center Fire ($44 billion), (2) the 1906 San Francisco Earthquake and Fire ($9 billion), (3) the 1871 Great Chicago Fire ($3.3 billion), and (4) the 1991 Oakland Hills Firestorm ($2.6 billion).
3. These are associated with the 2007 Southern California Fires (including "Witch" and "Harris") in San Diego County ($2.0 billion), the 2000 "Cerro Grande" Fire in Los Alamos, NM ($1.8 billion), the 2003 "Cedar" Fire in San Diego County, CA ($1.3 billion), and the 2003 "Old" Fire in San Bernardino Country, CA ($1.2 billion). National Fire Protection Association. 2013. Largest Loss Wildland Fires. NFPA Fire Incident Data Organization database. www.nfpa.org/research/reports-and-statistics/outdoor-fires/largest-loss-wildland-fires (accessed June 4, 2015).
4. These include 375,000 acres (Southern California Complex); 48,000 acres (Cerro Grande); 275,000 acres (Cedar Fire); and 91,000 acres (Old Fire). Cal Fire. 2014. Top 20 Most Damaging California Wildfires. Cal Fire Incident

Information. www.fire.ca.gov/communications/downloads/fact_sheets/Top20_Damaging.pdf (accessed June 4, 2015).

5. See for example, Alagona, P. 2013. *After the Grizzly: Endangered Species and the Politics of Place in California*. Berkeley: University of California Press.

6. National Wildfire Coordinating Group. 2014. Wildland Urban Interface Wildfire Mitigation Desk Reference Guide. PMS 051. August.

7. Sommers, W.T. 2008. The Emergence of the Wildland-Urban Interface Concept. *Forest History Today* (Fall), 12–18.

8. Davis, J. 1990. The Wildland-Urban Interface: Paradise or Battleground? *Journal of Forestry* 1, 26–31.

9. United States Department of Agriculture. 2013. Wildfire, Wildlands, and People: Understanding and Preparing for Wildfire in the Wildland-Urban Interface. USDA Rocky Mountain Research Station, January.

10. Cohen, J. 2008. The Wildland-Urban Interface Fire Problem: A Consequence of the Fire Exclusion Paradigm. *Forest History Today* (Fall), 20–26.

11. Fire Adapted Communities Coalition. 2014. Guide to Fire Adapted Communities.

12. Hewitt, K. 1997. *Regions of Risk: A Geographical Introduction to Disasters*. Essex, UK: Addison Wesley Longman; Hewitt, K. 1983. *Interpretations of Calamity*. London: Unwin Hyman; Liverman, D.M. 1990. Drought impacts in Mexico: Climate, Agriculture, Technology, and Land Tenure in Sonora and Puebla. *Annals of the Association of American Geographers* 80:1, 49–72; Wisner, B. 1993. Disaster Vulnerability: Scale, Power, and Daily Life. *GeoJournal* 30:2, 127–40.

13. Pelling, M. 2003. Toward a Political Ecology of Urban Environmental Risk. In *Political Ecology: An Integrative Approach to Geography and Environment-Development Studies*, ed. K.S. Zimmerer and T.J. Basset, 73–93. New York: Guildford; Oliver-Smith, A. 2002. Theorizing Disasters: Nature, Culture, Power. In *Culture and Catastrophe: The Anthropology of Disaster*, ed. S.M. Hoffman and A. Oliver-Smith. Santa Fe, NM: School of American Research Press; Orsi, J. 2004. *Hazardous Metropolis: Flooding and Urban Ecology in Los Angeles*. Berkeley: University of California Press; Cutter, S.L., Mitchell, J.T., and Scott, M.S. 2000. Revealing the Vulnerability of People and Places: A Case Study of Georgetown County, South Carolina. *Annals of the Association of American Geographers* 90:4, 713–37; Wisner, B., Blaikie, P., Cannon, T., and Davis, I. 2004. *At Risk: Natural Hazards, People's Vulnerability, and Disasters*, 2nd ed. London: Routledge; Mustafa, D. 1998. Structural Causes of Vulnerability to Flood Hazard in Pakistan. *Economic Geography* 74:3, 289–305; Dooling, S., and Simon, G. 2012. *Cities, Nature, Development: The Politics and Production of Urban Vulnerabilities*. Aldershot, UK: Ashgate; Collins, T. 2005. Households, Forests, and Fire Hazard Vulnerability in the American West: A Case Study of a California Community. *Global Environmental Change B: Environmental Haz-*

ards 6:1, 23–37; Collins, T. 2008. The Political Ecology of Hazard Vulnerability: Marginalization, Facilitation and the Production of Differential Risk to Urban Wildfires in Arizona's White Mountains. *Journal of Political Ecology* 15, 21–43; Collins, T. 2010. Marginalization, Facilitation, and the Production of Unequal Risk: The 2006 Paso del Norte Floods. *Antipode* 42:2, 258–88; Simon, G.L., and Dooling, S. 2013. Flame and Fortune in California: The Material and Political Dimensions of Vulnerability. *Global Environmental Change* 23:6, 1410–23; Simon, G.L. 2014. Vulnerability-in-Production: A Spatial History of Nature, Affluence, and Fire in Oakland, California. *Annals of the Association of American Geographers* 104:6, 1199–1221.

14. Mustafa, D. 2005. The Production of Urban Hazardscape in Pakistan: Modernity, Vulnerability and the Range of Choice. *Annals of the Association of American Geographers* 95:3, 566–86.

15. Adger, N. 2006. Vulnerability. *Global Environmental Change* 16:3, 268–81, 272.

16. Williams, R. 1975. *The Country and the City*. Oxford, UK: Oxford University Press; Gandy, M. 2002. *Concrete and Clay: Reworking Nature in New York City*. Cambridge, MA: MIT Press; Cronon, W. 1991. *Nature's Metropolis: Chicago and the Great West*. New York: Norton; Swyngedouw, E. 2004. *Social Power and the Urbanization of Water: Flows of Power*. Oxford, UK: Oxford University Press; Kaika, M. 2005. *City of Flows: Modernity, Nature, and the City*. New York: Routledge.

17. Braun, B. 2005. Writing a More-than-Human Urban Geography. *Progress in Human Geography* 29, 635–50, 642.

18. Wolch, J., Pincetl, S., and Pulido, L. 2001. Urban Nature and the Nature of Urbanism. In *From Chicago to L.A.: Making Sense of Urban Theory*, ed. M. Dear, 367–402. London: Sage; Keil, R., and Desfor, G. 2004. *Nature and the City: Making Environmental Policy in Toronto and Los Angeles*. Tucson: University of Arizona Press; Heynen, N., Kaika, M., and Swyngedouw, E. 2006. *In the Nature of Cities: Urban Political Ecology and the Politics of Urban Metabolism*. New York: Routledge; Dooling, S., Simon, G., and Yocom, K. 2006. Place-based Urban Ecology: A Century of Park Planning in Seattle. *Urban Ecosystems* 9:4, 299–321; Robbins, P. 2007. *Lawn People: How Grasses, Weeds, and Chemicals Make Us Who We Are*. Philadelphia: Temple University Press; Perkins, H. 2011. Gramsci in Green: Neoliberal Hegemony through Urban Forestry and the Potential for a Political Ecology of Praxis. *Geoforum* 42, 558–66; Muller, E., and Dooling, S. 2011. Sustainability and Vulnerability: Integrating Equity into Plans for Central City Redevelopment. *Journal of Urbanism* 4:3, 201–22; Swyngedouw, E. 2004. *Social Power and the Urbanization of Water: Flows of Power*. Oxford, UK: Oxford University Press; Simon, G.L. 2014. Vulnerability-in-Production: A Spatial History of Nature, Affluence, and Fire in Oakland, California. *Annals of the Association of American Geographers* 104:6, 1199–1221.

19. Harvey, D. 1973. *Social Justice and the City*. Baltimore: Johns Hopkins University Press; Harvey, D. 1996. *Justice, Nature and the Geography of Difference*. Oxford, UK: Blackwell.

20. Blaikie, P., Cannon, T., Davis, I., and Wisner, B. 1994. *At Risk: Natural Hazards, Peoples' Vulnerability and Disasters*. London: Routledge, 9. See also, Mustafa, D. 2005. The Production of Urban Hazardscape in Pakistan: Modernity, Vulnerability and the Range of Choice. *Annals of the Association of American Geographers* 95:3, 566–86, 566.

21. Watts, M. 1983. On the Poverty of Theory: Natural Hazards Research. In *Interpretation of Calamity: From the Viewpoint of Human Ecology*, ed. Kenneth Hewitt, 231–62. Boston: Allen & Unwin; Watts, M. 1983. *Silent Violence: Food, Famine, and Peasantry in Northern Nigeria*. Athens: University of Georgia Press.

22. Ray-Bennett, N.S. 2007. Environmental Disasters and Disastrous Policies: An Overview from India. *Social Policy and Administration* 41:4, 419–24, 420.

23. Davis, M. 1998. *Ecology of Fear: Los Angeles and the Imagination of Disaster*. New York: Metropolitan Books, 9. Here Davis juxtaposes this long history of development follies in the Los Angeles area against media portrayals depicting the city as a mere victim by unfortunate natural disasters. In so doing, these headlines naturalize what should otherwise be viewed as man-made disasters and conceal (and exonerate) the controversial inequities embedded in historical class- and race-based planning decisions.

24. Stengers, I. 2011. "Another Science Is Possible!" A Plea for Slow Science. Inaugural lecture, Chair Willy Calewaert 2011–12. Faculté de Philosophie et Lettres, Université Libre de Bruxelles. December 12.

25. Lave, R. 2012. Neoliberalism and the Production of Environmental Knowledge. *Environment and Society* 3, 19–38.

26. Brechin, G. 2006. *Imperial San Francisco: Urban Power, Earthly Ruin*. Berkeley: University of California Press, xxix (italics in original).

27. Pincetl, S., Rundel, P.W., De Blasio, J.C., Silver, D., Scott, T., Keeley, J.E., and Halsey, R. 2008. It's the Land Use Not the Fuels: Fires and Land Development in Southern California. *Real Estate Review* 37, 25–42.

CHAPTER 2: THE CHANGING AMERICAN WEST

1. Green, M. 2013. The Flammable West: Mega-Fires in the Age of Climate Change (with real-time fire map). *KQED News*, August 9.

2. Holthaus, E. 2016. We Need to Talk about Climate Change: Tragedies like the Fort McMurray Fire Make It More Important, Not Less. *Slate*, May 6.

3. Kolbert, E. 2016. Fort McMurray and the Fires of Climate Change. *New Yorker*, May 5.

4. Somer, W.T., Coloff, S.G., and Conard, S.G. 2011. Fire Regimes. In *Synthesis of Knowledge: Fire History and Climate Change*, ch. 3. Project 09-02-1-09. Boise, ID: Joint Fire Science Program.

5. Pyne, S.J. 1982. *Fire in America: A Cultural History of Wildland and Rural Fire*. Princeton, NJ: Princeton University Press.

6. Dennison, P.E., Brewer, S.C., Arnold, J.D., and Moritz, M.A. 2014. Large Wildfire Trends in the Western United States, 1984–2011. *Geophysical Research Letters* 41:8, 2928–33.

7. Flannigan, M.D., Krawchuk, M.A., de Groot, W.J., Wotton, B.M., and Gowman, L.M. 2009. Implications of Changing Climate for Global Wildland Fire. *International Journal of Wildland Fire* 18, 483–507.

8. Westerling, A.L., Hidalgo, H.G., Cayan, D.R., and Swetnam, T.W., 2006. Warming and Earlier Spring Increase Western US Forest Wildfire Activity. *Science* 313, 940–43.

9. International Association of Wildland Fire. 2013. www.iawfonline.org/pdf/WUI_Fact_Sheet_08012013.pdf (accessed May 25, 2015).

10. Cal Fire, 2013. State of California Wildfire Incident Information. http://cdfdata.fire.ca.gov/incidents/incidents_statsevents (accessed July 17, 2013).

11. Eichenseher, T. 2012. In Rocky Mountain Forests, More Fires and More People. *National Geographic*, June 28, 2012 (with permission from original author Michael Kodas).

12. Ingalsbee, T., 2010. *Getting Burned: A Taxpayer's Guide to Wildfire Suppression Costs*. Eugene, OR: Firefighters United for Safety, Ethics and Ecology.

13. International Association of Wildland Fire. 2013. www.iawfonline.org/pdf/WUI_Fact_Sheet_08012013.pdf (accessed May 25, 2015).

14. Botts, H., Jeffery, T., McCabe, S., Stueck, B., and Suhr, L. 2015. Wildfire Hazard Risk Report: Residential Wildfire Exposure Estimates for the Western United States. Irvine, CA: CoreLogic.

15. Pyne, S.J. 1982. *Fire in America: A Cultural History of Wildland and Rural Fire*. Princeton, NJ: Princeton University Press.

16. Davis, M. 1998. Let Malibu Burn: A Political History of the Fire Coast. *Radical Urban Theory*.

17. United States Census Bureau. 2016. www.census.gov/popclock/data_tables.php?component=growth (accessed April 19, 2016).

18. Albrecht, D.E. 2007. The Changing West: A Regional Overview. In *Population Brief: Trends in the Western U.S.* Logan, UT: Western Rural Development Center.

19. Federal Emergency Management Agency. 2002. Fires in the Wildland/Urban Interface. U.S. Fire Administration, Topical Fire Research Series 2(16).

20. Hammer, R.B., Radeloff, V.C., Freid, J.S., and Stewart, S.I. 2007. Wildland-Urban Interface Housing Growth during the 1990s in California, Oregon, and Washington. *International Journal of Wildland Fire* 16, 255–65.

21. International Association of Wildland Fire. 2013. www.iawfonline.org/pdf/WUI_Fact_Sheet_08012013.pdf (accessed May 25, 2015).

22. Theobald, D.M., and Romme, W.H. 2007. Expansion of the US Wildland-Urban Interface. *Landscape and Urban Planning* 83, 340–54.

23. International Association of Wildland Fire. 2013. www.iawfonline.org/pdf/WUI_Fact_Sheet_08012013.pdf (accessed May 25, 2015).

24. Gorte, R. 2013. The Rising Cost of Wildfire Protection. Bozeman, MT: Headwaters Economics. http://headwaterseconomics.org/wildfire/fire-costs-background/; Gude, P., Rasker, R., and van den Noort, J. 2008. Potential for Future Development on Fire-prone Lands. *Journal of Forestry* 106, 198–205.

25. U.S. Forest Service. n.d. National Forests on the Edge: Development Pressures on America's National Forests and Grasslands. www.fs.fed.us/openspace/fote/national_forests_on_the_edge.html (accessed May 25, 2015).

26. Moritz, M.A., Batllori, E., Bradstock, R.A., Gill, A.M., Handmer, J., Hessburg, P.F., Leonard, J., McCaffrey, S., Odion, D.C., Schoennagel, T., and Syphard, A.D. 2014. Learning to Coexist with Wildfire. *Nature* 515, 58–66.

27. Sibold, J.S., Veblen, T.V., and Gonzalez, M.E. 2006. Spatial and Temporal Variation in Historic Fire Regimes in Subalpine Forests across the Colorado Front Range in Rocky Mountain National Park, Colorado, USA. *Journal of Biogeography* 32, 631–47.

28. It is important to note the role of El Niño Southern Oscillation (ENSO) events and their impact on regional climates, including precipitation, temperature, and aridity over multiyear periods. Along with changes to the global climate these shifting regional oceanic/atmospheric processes significantly alter fire activity.

29. See Peterson, T.C., Heim Jr., R.R., Hirsch, R. Kaiser, D.P., Brooks, H., Diffenbaugh, N.S., Dole, R.M., Giovannettone, J.P., Guirguis, K., Karl, T.R., Katz, R.W., Kunkel, K., Lettenmair, D., McCabe, G.J., Paciorek, C.J., Ryberg, K.R., Schubert, S., Silva, V.B.S., Stewart, B.G., Vecchia, A.V., Villiarini, G., Vose, R.S., Walsh, J., Wehner, M., Wolock, D., Wolter, K., Woodhouse, G.A., and Wuebbles, D. 2013. Monitoring and Understanding Changes in Heat Waves, Cold Waves, Floods and Droughts in the United States: State of Knowledge. *Bulletin of the American Meteorological Society* 94:6, 821–34.

30. Romero-Lankao, P., Smith, J.B., Davidson, D.J., Diffenbaugh, N.S., Kinney, P.L., Kirshen, P., Kovacs, P., and Villers Ruiz, L. 2014. North America. In *Climate Change 2014: Impacts, Adaptation, and Vulnerability. Part B: Regional Aspects. Contribution of Working Group II to the Fifth Assessment Report of the Intergovernmental Panel on Climate Change*, ed. V.R. Barros, C.B. Field, D.J. Dokken, M.D. Mastrandrea, K.J. Mach, T.E. Bilir, M. Chatterjee, K.L. Ebi,

Y.O. Estrada, R.C. Genova, B. Girma, E.S. Kissel, A.N. Levy, S. MacCracken, P.R. Mastrandrea, and L.L. White, 1439–98. Cambridge, UK/New York: Cambridge University Press.

31. See also, Westerling, A.L., Hidalgo, H.G., Cayan, D.R., and Swetnam, T.W. 2006. Warming and Earlier Spring Increase Western US Forest Wildfire Activity. *Science* 313, 940–43; Williamson, T.B., Colombo, S.J., Duinker, P.N., Gray, P.A., Hennessey, R.J., Houle, D., Johnston, M.H., Ogden, A.E., and Spittlehouse, D.L. 2009. *Climate Change and Canada's Forests: From Impacts to Adaptation*. Sustainable Forest Management Network and Natural Resources Canada, Canadian Forest Service, Northern Forestry Centre, Edmonton, AB, 104; Holden, Z.A., Morgan, P., Crimmins, M.A., Steinhorst, R.K., and Smith, A.M. 2007. Fire Season Precipitation Variability Influences Fire Extent and Severity in a Large Southwestern Wilderness Area, United States. *Geophysical Research Letters* 34:16, L16708, doi:10.1029/2007GL030804.

32. Fagan, K. 2014. "Unprecedented" Winter Wildfires in Far Northern California. *San Francisco Chronicle*, January 6.

33. Union of Concerned Scientists. 2013. Is Global Warming Fueling Increased Wildfire Risks? www.ucsusa.org/global_warming/science_and_impacts/impacts/global-warming-and-wildfire.html#.VWyS5lwrjdk.

34. Marlon, J.R., Bartlein, P.J., Gavin, D.G., Long, C.J., Anderson, R.S., Briles, C.E., Brown, K.J., Colombarolig, D., Halletth, D.J., Poweri, M.J., Scharfj, E.A., and Walsh, M.K. 2012. Long-term Perspective on Wildfires in the Western USA. *Proceedings of the National Academy of Sciences* 109, E535–E543.

CHAPTER 3: TRAILBLAZING

1. The 1915 Claremont Hotel was actually a rebuild project of a preexisting construction that burned down several years earlier.

2. Although more southerly neighborhoods containing most of the area's redwood stands did not burn during the Tunnel Fire, those communities certainly remain very susceptible to future firestorms. Hence the connection between productive logging and household risk is germane to the entire area and not simply the 1991 impacted fire zone.

3. Bagwell, B. 1982. *Oakland: The Story of a City*. Novato, CA: Presidio Press, 15.

4. Slack, G. 2004. In the Shadow of Giants: The Redwoods of the East Bay Hills. *Bay Nature*, July 1. https://baynature.org/articles/in-the-shadow-of-giants/.

5. Bagwell, B. 1982. *Oakland: The Story of a City*. Novato, CA: Presidio Press.

6. City of Oakland. 1996. Open Space Conservation and Recreation (OSCAR): An Element of the General Plan.

7. These include roads such as Redwood Road and Park Boulevard.
8. These include present-day Claremont and Thornhill roads.
9. Bagwell, B. 1982. *Oakland: The Story of a City.* Novato, CA: Presidio Press, 18.
10. Burgess, S.D. 1951. "The Forgotten Redwoods of the East Bay." *California Historical Society Quarterly* 30(March), 10.
11. Nowak, D.J. 1993. Historical Vegetation Change in Oakland and Its Implication for Urban Forest Management. *Journal of Arboriculture* 19:5.
12. Tyrell, I. 1999. *True Garden of the Gods: California-Australian Environmental Reform 1860–1930.* Berkeley: University of California Press.
13. Nowak, D.J. 1993. Historical Vegetation Change in Oakland and Its Implication for Urban Forest Management. *Journal of Arboriculture* 19:5.
14. The Realty Syndicate. 1911. Realty Syndicate. In *Greater Oakland*, ed. E. Blake. Oakland, CA: Pacific Publishing, 269.
15. *Oakland Tribune.* 1923. Oakland Hills' Forest Mantle All Handmade. July 15, p. 9.
16. O'Brien, B. 2005. Ubiquitous Eucalyptus: How an Aussie Got Naturalized. *Bay Nature* (July–September), 1–6, 2.
17. City of Oakland. 1998. Envision Oakland: An Element of the General Plan, 205.
18. Bagwell, B. 1982. *Oakland: The Story of a City.* Novato, CA: Presidio Press.
19. The Realty Syndicate. 1911. Realty Syndicate. In *Greater Oakland*, ed. E. Blake. Oakland, CA: Pacific Publishing, 261.
20. City of Oakland. 1986. North Oakland Hill Area Specific Plan. Oakland City Planning Department, 112. See also Farmer, J. 2013. *Trees in Paradise: A California History.* New York: Norton.
21. Federal Emergency Management Agency. 1992. The East Bay Hills Fire Oakland-Berkeley, California. U.S. Fire Administration/Technical Report Series USFA-TR-060. Washington, DC: FEMA, 7.
22. LSA Associates Inc. 2009. Wildfire Hazard Reduction and Resource Management Plan for the East Bay Regional Park District. Draft Report, July 20, p. 13.
23. Adger, N. 2006. Vulnerability. *Global Environmental Change* 16:3, 268–81, 270.

CHAPTER 4: SETTING THE STAGE FOR DISASTER

1. For a discussion of earlier, post–Gold Rush suburbanization in the San Francisco Bay Area (including the role of Alameda County and the City of Oakland) see Walker, R. 2004. Industry Builds Out the City: The Suburbanization of

Manufacturing in the San Francisco Bay Area 1850–1940. In *The Manufactured Metropolis*, ed. Robert Lewis, 92–123. Philadelphia: Temple University Press.

2. Self, R.O. 2003. *American Babylon: Race and the Struggle for Postwar Oakland.* Princeton, NJ: Princeton University Press.

3. Silva, F., and Barbour, E. 1999. *The State-Local Fiscal Relationship in California: A Changing Balance of Power.* San Francisco: Public Policy Institute of California.

4. Brownlee, E. 2003. California Taxes: Historical Roots, and the Property Tax: Will It Survive? Brief on Proposition 13 prepared by Assembly Committees on Local Government and Revenue and Taxation, IGS.

5. Ross, J. 2009. Proposition 13 Thirty Years Later: What Has It Meant for Governance and Public Services? In *After the Tax Revolt: Proposition 13 Turns 30*, ed. J. Citrin and I. Martin, ch. 9, 135–40. Berkeley, CA: Public Policy Press.

6. Chapman, J.I. 1998. Proposition 13: Some Unintended Consequences. San Francisco: Public Policy Institute of California.

7. Self, R.O. 2003. *American Babylon: Race and the Struggle for Postwar Oakland.* Princeton, NJ: Princeton University Press.

8. Federal Emergency Management Agency. 1992. The East Bay Hills Fire Oakland-Berkeley, California. U.S. Fire Administration/Technical Report Series USFA-TR-060, Washington, DC: FEMA, 2.

9. Ibid., 50.

10. City of Oakland. 1996. Open Space Conservation and Recreation (OSCAR): An Element of the General Plan, 38.

11. City of Oakland. 1992. The Oakland Fire Department's Response to the Office of Emergency Service Report: The East Bay Hills Fire of October 20, 1991. Office of the Fire Chief.

12. Serna, J. 2014. California Drought Brings "Unprecedented" Fire Danger. *Los Angeles Times*, January 18.

13. State of California. 2001. Oral History Interview with Tom Bates. California State Archives. State Government Oral History Program. By L. McGarrigle, Regional Oral History Office, University of California.

14. Hoffer, P. 2006. *Seven Fires: The Urban Infernos That Reshaped America.* New York: Public Affairs.

15. Pincetl, S., Rundel, P.W., De Blasio, J.C., Silver, D., Scott, T., Keeley, J.E., and Halsey, R. 2008. It's the Land Use Not the Fuels: Fires and Land Development in Southern California. *Real Estate Review* 37, 25–42.

16. Chapman, J.I. 1998. Proposition 13: Some Unintended Consequences. San Francisco: Public Policy Institute of California.

17. The year 2012 is used here as an example of analysis. Similar numbers can be generated from other annual tax roll assessments.

18. It is important to keep in mind that in California local governments receive nearly all their funding through property taxes. Hence the dramatic impacts of Proposition 13 on local fire departments.

19. City of Oakland. 1986. North Oakland Hill Area Specific Plan. Oakland City Planning Department, 122.

20. City of Oakland. 1992. The Oakland Fire Department's Response to the Office of Emergency Service Report: The East Bay Hills Fire of October 20, 1991. Office of the Fire Chief, 122.

21. Stern, S. 1991. Firestorm Heats Up Debate over Development in Hills. *Oakland Tribune.*

22. Ibid.

23. Kemp, R.L. 1980. *Coping with Proposition 13.* Lexington, MA: Lexington Books.

24. See for example Davis, M. 1998. *Ecology of Fear: Los Angeles and the Imagination of Disaster.* New York: Metropolitan Books; and, Wolch, J., Pincetl, S., and Pulido, L. 2001. Urban Nature and the Nature of Urbanism. In *From Chicago to L.A.: Making Sense of Urban Theory,* ed. M. Dear, 367–402. London: Sage.

25. State of California. 1991. The 1991 East Bay Firestorm. California Legislative State Senate. R. Torres, Chairman. Senate Committee on Toxics and Public Safety Management.

26. Household perceptions of actual fire risk may also have been reduced as a result of a persuasive scale construction deployed by the antitax movement. Through effective campaign strategies tax opponents targeted suburban residents by suggesting that local property taxes are collected and redistributed from suburban areas into the metropolitan core. The ideologically driven nature of the movement masked the actual—and much more localized (i.e., intracity)—spatial extent of the property tax collection and redistribution catchment.

27. Rodrigue, C.M. 1993. Home with a View: Chaparral Fire Hazard and the Social Geographies of Risk and Vulnerability. *California Geographer* 33, 105–18.

28. Ross, J. 2009. Proposition 13 Thirty Years Later: What Has It Meant for Governance and Public Services? In *After the Tax Revolt: Proposition 13 Turns 30,* ed. J. Citrin and I. Martin, ch. 9, 135–40. Berkeley, CA: Public Policy Press.

29. Greenwood, D., and Brown, T. 2003. *An Overview of Colorado's State and Local Tax Structures.* Colorado Springs: Center for Colorado Policy Studies. See also Gale, W., Houser, S., and Scholz, J.K. 1996. Distributional Effects of Fundamental Tax Reform. In *Economic Effects of Fundamental Tax Reform,* ed. H. Aaron and W. Gale, ch. 8. Washington, DC: Brookings Institution.

30. Gamage, D. 2009. Coping through California's Budget Crises in Light of Proposition 13 and California's Fiscal Constitution, Proposition 13 at 30. In *After*

the Tax Revolt: Proposition 13 Turns 30, ed. J. Citrin and I. Martin, ch. 4, 51–67. Berkeley, CA: Public Policy Press.

31. McClintock, N. 2011. From Industrial Garden to Food Desert: Demarcated Devaluation in the Flatlands of Oakland, California. In *Cultivating Food Justice: Race, Class and Sustainability,* ed. A. Alkon and J. Agyeman, 89–120. Cambridge, MA: MIT Press.

32. Laymance Real Estate Company. 1911. What the Development of the Foothills Has Meant to Oakland's Growth. In *Greater Oakland,* ed. E. Blake, 261–68. Oakland, CA: Pacific Publishing, 266.

33. Ibid., 261.

34. Harvey, D. 1996. Justice, Nature and the Geography of Difference. Oxford, UK: Blackwell, 183.

35. Walker, R. 1995. Landscape and City Life: Four Ecologies of Residence in the San Francisco Bay Area. *Ecumene* 2:1, 33–64, 33.

36. Laymance Real Estate Company. 1911. The Protection Rockridge Affords. Oakland, CA: Laymance Real Estate Company, 6.

37. Citrin, J., and Martin, I., eds. 2009. *After the Tax Revolt: Proposition 13 Turns 30.* Berkeley, CA: Public Policy Press, 5.

CHAPTER 5: WHO'S VULNERABLE?

1. Thanks to Christine Eriksen and also members of the UCLA Department of Geography for raising this distinction and also for pressing other key insights on the meaning and nature of fire vulnerability in the context of affluence.

2. Sovern, D. 2011. Families Still Struggling with Loss 20 Years after Oakland Firestorm. *CBS SF Bay Area,* October 19. http://sanfrancisco.cbslocal.com/2011/10/19/families-still-struggling-with-loss-20-years-after-oakland-firestorm/.

3. Johnson, S. 2011. The Destruction Was Just the First Wave of Pain for Many Survivors of the Oakland Hills Fire. *Inside BayArea,* October 19. www.insidebayarea.com/oaklandtribune/localnews/ci_19148957.

4. United Policyholders. www.uphelp.org/about/mission (accessed July 6, 2015).

CHAPTER 6: SMOKE SCREEN

1. To be sure, the issue of fire and fire risk will always remain political and politicized in some form. However I argue here that it is the role of the Incendiary with its associated controversies in particular that gets diminished and obscured within these mainstream discursive representations.

2. Federal Emergency Management Agency. 1992. The East Bay Hills Fire Oakland-Berkeley, California. U.S. Fire Administration/Technical Report Series USFA-TR-060, Washington, DC: FEMA.

3. Ewell, P.L. 1995. The Oakland-Berkeley Hills Fire of 1991. USDA Forest Service Gen. Tech. Rep. PSW-GTR-158.

4. National Fire Protection Association. 1992. The Oakland/Berkeley Hill Fire. National Wildland/Urban Interface Fire Protection Initiative.

5. Beck, U. 1992. *Risk Society: Towards a New Modernity*. London: Sage.

6. It is understood that the term *Anthropocene* is fraught with conceptual shortcoming. These include the tendency for the concept to flatten responsibility across temporal, geographical, and socioeconomic categories (i.e., its not "humanity's" doing but rather the result of particular individuals, regions, and economies) and also the opaque if not absent indication of what new, emergent, and transformed socioecological conditions are acceptable (i.e., what new baselines are we managing for or against?). Moreover the precise beginning of the Anthropocene is still open for debate. For some it begins with the Industrial Revolution, while for others its origins can be traced back over 8,000 years (thus considerably overlapping with the Holocene) to early periods of large land modifications and agricultural intensification.

7. Of course many fires were not absent of human influence, as native populations and early settlers left a distinct imprint on the environment by altering landscapes and fire regimes in a myriad of ways.

8. Massey, N. 2012. Colorado Fire Follows in Pine Beetles' Tracks. *Scientific American, ClimateWire*, June 20.

9. Other factors however, such as the impacts of beetle kill on fire intensity, are not as clear. The jury is still out for example on whether infested forests present a higher risk to homeowners and firefighters when confronted by a major fire event.

10. Pielke, R.A. 2002. Policy, Politics and Perspective. *Nature* 416:368.

11. Jolly, W.M., Cochrane, M.A., Freeborn, P.H., Holden, Z.A., Brown, T.J., Williamson, G.J., and Bowman, D.M. 2015. Climate-induced Variations in Global Wildfire Danger from 1979 to 2013. *Nature Communications* 6.

12. Casey, M. 2015. Are Massive Wildfires the New Normal? *CBS News*, July 28. www.cbsnews.com/news/massive-wildfires-climate-change-the-new-normal/.

13. Gaillard, C., Glantz, M., Kelman, I., Wisner, B., Delica-Willison, Z., and Keim, M. 2014. Taking the "Naturalness" out of Natural Disaster (Again). *Natural Hazards Observer*, January, 1, 14–16.

14. Ironically this announcement came just as California was reeling from intense rainfall and flooding across the state. These El Niño Southern Oscillation (ENSO)-induced storms point to the complexity of weather patterns that influence fire activity. Many at Cal Fire see this rain as a mixed blessing because

although more snow is falling in the mountains and the threat of winter fires is reduced, the long-term result of this moisture is more grasses and underbrush that could eventually become flammable tinder if typical arid conditions pick up again in summer and fall.

15. Mai-Duc, C. 2016. Brown's Budget Earmarks Big Money for Natural Disasters. *Los Angeles Times*, January 8.

16. Simon, R. 1989. City, County Take Different Paths on Wood Roofs Issue. *Los Angeles Times*, July 7.

17. Ibid.

18. Hull, T. 1991. Oakland Fire Spurs Effort to Ban Untreated Wood Roofing in California. *San Francisco Examiner*, November 2.

19. Federal Emergency Management Agency. 1992. The East Bay Hills Fire Oakland-Berkeley, California. U.S. Fire Administration/Technical Report Series USFA-TR-060, Washington, DC: FEMA.

20. Friederici, P. 2009. California Prepares for the Next Burn: Officials—and Homeowners—Start to Accept the Inevitability of Wildfire. *High Country News*, June 1, p. 20.

21. Hull, T. 1991. Oakland Fire Spurs Effort to Ban Untreated Wood Roofing in California. *San Francisco Examiner*, November 2.

22. National Fire Protection Association. 1992. The Oakland/Berkeley Hill Fire. National Wildland/Urban Interface Fire Protection Initiative.

23. Muir, F. 1989. Wood-Roof Ban Sent to Bradley on 11–1 Vote: Industry in Near-Panic. *Los Angeles Times*, July 13.

24. Frammolino, R. 1991. Tragedy Renews Calls for Ban on Wooden Shingles: Prevention: Some communities require that roofing materials be fire-resistant, but measures that would have imposed statewide standards have not been able to get past opposition. *Los Angeles Times*, October 22.

25. Simon, G. 2014. Vulnerability-in-Production: A Spatial History of Nature, Affluence and Fire in Oakland, California. *Annals of the Association of American Geographers* 104:6, 1199–1221.

CHAPTER 7: DEBATES OF DISTRACTION

1. Federal funds are being distributed to three entities in charge of overseeing and implementing the project: the University of California, Berkeley; the City of Oakland; and the East Bay Regional Parks District.

2. Orenstein, N. 2015. Hills Group Sues FEMA over Plan to Cut Down Trees. *Berkeleyside*, March 23.

3. Because home developments in the Oakland Hills are already locked in place, it may seem that efforts to question the Incendiary are both too late and unnecessary; at this point we need to emphasize adaptation measures in

decision making. However these arguments over vegetation management occur in many other new and emerging high-risk locations around the West. The battle over eucalypts is thus symbolic of other debates of distraction that come to occupy and saturate the discursive arena of dispute in fire-prone landscapes in the region.

4. Taylor, T. 2013. UC Berkeley Expert Talks about Hillside Tree Removal Plan. *Berkeleyside*, June 11.

5. United States Department of Agriculture. 1995. Comparison of Fuel Load, Structural Characteristics and Infrastructure before and after the Oakland Hills "Tunnel Fire." USDA Forest Service Gen. Tech. Rep. PSW-GTR-158.

6. East Bay Regional Park District. n.d. Background Report: The East Bay Hills Wildfire Problem Statement. www.ebparks.org/about/stewardship fuelsplan/bg_report (accessed May 11, 2015).

7. See for example, Hobbs, R.J., et al. 2006. Novel Ecosystems: Theoretical and Management Aspects of the New Ecological World Order. *Global Ecology and Biogeography* 15:1, 1–7; and Marris, E. 2013. *Rambunctious Garden: Saving Nature in a Post-wild World*. New York: Bloomsbury USA.

8. Lochner, T. 2014. Oakland-Berkeley Hills Groups at Odds over Eucalyptus Fire Hazard Abatement Tactics. *Contra Costa Times*, October 1.

9. Dornheim, R. 2011. Oakland Firestorm Anniversary: The Fight over Eucalyptus Trees Continues. *KQED News*, October 19.

10. Alagona, P. 2013. *After the Grizzly: Endangered Species and the Politics of Place in California*. Berkeley: University of California Press.

11. A $35 reduction is given to property owners falling within the boundaries of a local fire protection agency.

12. Howard Jarvis Taxpayers Association. 2012. http://firetaxprotest.org/wp-content/uploads/2012/10/PR1003121.pdf (accessed August 22, 2015).

13. The nonpartisan Office of Legislative Council ruled that it qualifies as a fee because revenues are earmarked for specific state services.

14. Howard Jarvis Taxpayers Association. 2012. http://firetaxprotest.org/wp-content/uploads/2012/10/PR1003121.pdf (accessed August 22, 2015).

15. Buchanan, W. 2012. California Fire Fee Likely to Kindle Ire. *San Francisco Chronicle*. www.sfgate.com/news/article/California-fire-fee-likely-to-kindle-ire-3777487.php (accessed August 22, 2015).

CHAPTER 8: DISPATCHES FROM THE FIELD

1. Federal Emergency Management Agency. 1992. The East Bay Hills Fire Oakland-Berkeley, California. U.S. Fire Administration/Technical Report Series USFA-TR-060, Washington, DC: FEMA.

2. Ewell, P.L. 1995. The Oakland-Berkeley Hills Fire of 1991. USDA Forest Service Gen. Tech. Rep. PSW-GTR-158.

3. Federal Emergency Management Agency. 1992. The East Bay Hills Fire Oakland-Berkeley, California. U.S. Fire Administration/Technical Report Series USFA-TR-060, Washington, DC: FEMA.

4. Water Distribution Planning Division. 1992. Analysis of Broadway Terrace Pressure Zone B4Aa. East Bay Municipal Utility District, Oakland, CA.

5. Scawthorn, C., Eiginger, J.M., and Schiff, A.J., eds. 2005. *Fire Following Earthquake*. Reston, VA: American Society of Civil Engineers.

6. Blonski, K.S., Morales, M.E., and Morales, T.J. 2002. *Proceedings of the California's 2001 Wildfire Conference, Ten Years after the East Bay Hills Fire*, October 10–12, 2001, Oakland, CA. Technical Report 35.01.462. Richmond, CA: University of California Forest Products Laboratory.

7. Blonski, K.S., Miller, C., and Rice, C.L. 2011. History: Tunnel Fire, 20 Years After. *Wildfire World* (International Association of Wildfire), September 23.

8. Ibid.

9. Federal Emergency Management Agency. 1992. The East Bay Hills Fire Oakland-Berkeley, California. U.S. Fire Administration/Technical Report Series USFA-TR-060, Washington, DC: FEMA.

10. Blonski, K.S., Miller, C., and Rice, C.L. 2011. History: Tunnel Fire, 20 Years After. *Wildfire World* (International Association of Wildfire), September 23.

11. Blonski, K.S., Morales, M.E., and Morales, T.J. 2002. *Proceedings of the California's 2001 Wildfire Conference, Ten Years after the East Bay Hills Fire*, October 10–12, 2001, Oakland, CA. Technical Report 35.01.462. Richmond, CA: University of California Forest Products Laboratory.

CHAPTER 9: OUT OF THE ASHES

1. Davis, M. 1998. *Ecology of Fear: Los Angeles and the Imagination of Disaster*. New York: Metropolitan Books, 108.

2. These data were compiled by Ryan Maye Handy with the *Colorado Springs Gazette*. A synopsis can be found at Handy, R.M. 2014. Waldo Canyon Fire: Two Years Later, a Neighborhood Is Reborn. *Colorado Springs Gazette*, June 23. http://gazette.com/waldo-canyon-fire-two-years-later-a-neighborhood-is-reborn/article/1521839#ChUhyFfEPhMb5V6L.99.

3. WildfireX. 2015. www.wildfirex.com/private-firefighting/ (accessed March 30, 2015). Data based on a 2012 study conducted by market research firm IBISWorld.

4. Yoshino, K. 2007. Another Way the Rich Are Different: "Concierge-Level" Fire Protection. *Los Angeles Times*, October 26. www.latimes.com/business/la-fi-richfire26oct26-story.html#page = 1.

5. Ibid.

6. www.geltechsolutions.com/fireice/products.aspx (accessed April 1, 2015).

7. Department of Environment and Conservation, Western Australia. 2013. Fire Fighting Foams with Perfluorochemicals—Environmental Review. June.

8. www.consumerfireproducts.com/foam-gels-and-fire-retardant.html (accessed April 9, 2015).

9. Yoshino, K. 2007. Another Way the Rich Are Different: "Concierge-Level" Fire Protection. *Los Angeles Times*, October 26. www.latimes.com/business/la-fi-richfire26oct26-story.html#page = 1.

10. Handy, R. 2014. Wildfires Fan Growth of Private Firefighting Crews. *The Coloradoan*, May 31. www.coloradoan.com/story/money/2014/06/01/wildfires-fan-growth-private-firefighting-crews/9744857/.

CONCLUSION

1. International Association of Wildland Fire. 2013. Yarnell Hill Fire Serious Accident Investigation Report. September 23.

2. National Fire Protection Association. 2015. Deadliest Incidents Resulting in the Deaths of 8 or More Firefighters. www.nfpa.org/research/reports-and-statistics/the-fire-service/fatalities-and-injuries/deadliest-incidents-resulting-in-the-deaths-of-8-or-more-firefighters (accessed July 13, 2015).

3. www.hillsemergencyforum.org/mission-goals.html.

4. Stephens, S.L., Moghaddas, J.J., Ediminster, C., Fiedler, C.E., Hasse, S., Harrington, M., Keeley, J.E., McIver, J.D., Metlen, K., Skinner, C.N., and Youngblood, A. 2009. Fire Treatment Effects on Vegetation Structure, Fuels, and Potential Fire Severity in Western U.S. Forests. *Ecological Applications* 19, 305–20.

5. Pyne, S.J. 1982. *Fire in America: A Cultural History of Wildland and Rural Fire*. Princeton, NJ: Princeton University Press.

6. Headwaters Economics. 2009. *Solutions to the Rising Costs of Fighting Fires in the Wildland-Urban Interface*, updated ver. December. Bozeman, MT: Headwaters Economics.

7. Ibid.

8. Friederici P. 2009. California Prepares for the Next Burn: Officials—and Homeowners—Start to Accept the Inevitability of Wildfire. *High Country News*, June 1.

9. Troy, A., and Romm, J. 2006. An Assessment of the 1998 California Natural Hazard Disclosure Law (AB 1195). California Policy Research Center, University of California, Berkeley.

10. Headwaters Economics. 2009. *Solutions to the Rising Costs of Fighting Fires in the Wildland-Urban Interface,* updated ver. December. Bozeman, MT: Headwaters Economics.

11. Albrecht, D.E. 2007. The Changing West: A Regional Overview. In *Population Brief: Trends in the Western U.S.* Logan, UT: Western Rural Development Center.

12. U.S. Fire Administration; Federal Emergency Management Agency; Department of Homeland Security. 2002. Fires in the Wildland/Urban Interface. *Topical Fire Research Series* 2:16, 1–4. www.usfa.dhs.gov/downloads/pdf/tfrs/v2i16-508.pdf (accessed May 13, 2009).

References

Adger, N. 2006. Vulnerability. *Global Environmental Change* 16:3, 268–81.
Alagona, P. 2013. *After the Grizzly: Endangered Species and the Politics of Place in California.* Berkeley: University of California Press.
Albrecht, D.E. 2007. The Changing West: A Regional Overview. In *Population Brief: Trends in the Western U.S.* Logan, UT: Western Rural Development Center.
Bagwell, B. 1982. *Oakland: The Story of a City.* Novato, CA: Presidio Press.
Beck, U. 1992. *Risk Society: Towards a New Modernity.* London: Sage.
Blaikie, P., Cannon, T., Davis, I., and Wisner, B. 1994. *At Risk: Natural Hazards, Peoples' Vulnerability and Disasters.* London: Routledge.
Blonski, K.S., Miller, C., and Rice, C.L. 2011. History: Tunnel Fire, 20 Years After. *Wildfire World* (International Association of Wildfire), September 23.
Blonski, K.S., Morales, M.E., and Morales, T.J. 2002. *Proceedings of the California's 2001 Wildfire Conference, Ten Years after the East Bay Hills Fire,* October 10–12, 2001, Oakland, CA. Technical Report 35.01.462. Richmond, CA: University of California Forest Products Laboratory.
Botts, H., Jeffery, T., McCabe, S., Stueck, B., and Suhr, L. 2015. Wildfire Hazard Risk Report: Residential Wildfire Exposure Estimates for the Western United States. Irvine, CA: CoreLogic.
Braun, B. 2005. Writing a More-than-Human Urban Geography. *Progress in Human Geography* 29, 635–50.

Brechin, G. 2006. *Imperial San Francisco: Urban Power, Earthly Ruin.* Berkeley: University of California Press.

Brownlee, E. 2003. California Taxes: Historical Roots, and the Property Tax: Will It Survive? Brief on Proposition 13 prepared by Assembly Committees on Local Government and Revenue and Taxation, IGS.

Buchanan, W. 2012. California Fire Fee Likely to Kindle Ire. *San Francisco Chronicle.* www.sfgate.com/news/article/California-fire-fee-likely-to-kindle-ire-3777487.php (accessed August 22, 2015).

Burgess, S.D. 1951. "The Forgotten Redwoods of the East Bay." *California Historical Society Quarterly* 30(March), 10.

Cal Fire. 2013. State of California Wildfire Incident Information. http://cdfdata.fire.ca.gov/incidents/incidents_statsevents (accessed July 17, 2013).

———. 2014. Top 20 Most Damaging California Wildfires. CalFire Incident Information. www.fire.ca.gov/communications/downloads/fact_sheets/Top20_Damaging.pdf (accessed June 4, 2015).

Casey, M. 2015. Are Massive Wildfires the New Normal? *CBS News,* July 28. www.cbsnews.com/news/massive-wildfires-climate-change-the-new-normal/.

Chapman, J.I. 1998. *Proposition 13: Some Unintended Consequences.* San Francisco: Public Policy Institute of California.

Citrin, J., and Martin, I., eds. 2009. *After the Tax Revolt: Proposition 13 Turns 30.* Berkeley, CA: Public Policy Press.

City of Oakland. 1986. North Oakland Hill Area Specific Plan. Oakland City Planning Department.

———. 1992. The Oakland Fire Department's Response to the Office of Emergency Service Report: The East Bay Hills Fire of October 20, 1991. Office of the Fire Chief.

———. 1996. Open Space Conservation and Recreation (OSCAR): An Element of the General Plan.

———. 1998. Envision Oakland: An Element of the General Plan.

Cohen, J. 2008. The Wildland-Urban Interface Fire Problem: A Consequence of the Fire Exclusion Paradigm. *Forest History Today* (Fall), 20–26.

Collins, T. 2005. Households, Forests, and Fire Hazard Vulnerability in the American West: A Case Study of a California Community. *Global Environmental Change B: Environmental Hazards* 6:1, 23–37.

———. 2008. The Political Ecology of Hazard Vulnerability: Marginalization, Facilitation and the Production of Differential Risk to Urban Wildfires in Arizona's White Mountains. *Journal of Political Ecology* 15, 21–43.

———. 2010. Marginalization, Facilitation, and the Production of Unequal Risk: The 2006 Paso del Norte Floods. *Antipode* 42:2, 258–88.

Cronon, W. 1991. *Nature's Metropolis: Chicago and the Great West.* New York: Norton.

Cutter, S.L., Mitchell, J.T., and Scott, M.S. 2000. Revealing the Vulnerability of People and Places: A Case Study of Georgetown County, South Carolina. *Annals of the Association of American Geographers* 90:4, 713–37.

Davis, J. 1990. The Wildland-Urban Interface: Paradise or Battleground? *Journal of Forestry* 1, 26–31.

Davis, M. 1998. *Ecology of Fear: Los Angeles and the Imagination of Disaster.* New York: Metropolitan Books.

———. 1998. Let Malibu Burn: A Political History of the Fire Coast. *Radical Urban Theory.*

Dennison, P.E., Brewer, S.C., Arnold, J.D., and Moritz, M.A. 2014. Large Wildfire Trends in the Western United States, 1984–2011. *Geophysical Research Letters* 41:8, 2928–33.

Department of Environment and Conservation, Western Australia. 2013. Fire Fighting Foams with Perfluorochemicals—Environmental Review. June.

Dooling, S., and Simon, G. 2012. *Cities, Nature, Development: The Politics and Production of Urban Vulnerabilities.* Aldershot, UK: Ashgate.

Dooling, S., Simon, G., and Yocom, K. 2006. Place-based Urban Ecology: A Century of Park Planning in Seattle. *Urban Ecosystems* 9:4, 299–321.

Dornheim, R. 2011. Oakland Firestorm Anniversary: The Fight over Eucalyptus Trees Continues. *KQED News*, October 19.

East Bay Regional Park District. n.d. Background Report: The East Bay Hills Wildfire Problem Statement. www.ebparks.org/about/stewardship/fuelsplan/bg_report (accessed May 11, 2015).

Eichenseher, T. 2012. In Rocky Mountain Forests, More Fires and More People. *National Geographic*, June 28, 2012 (with permission from original author Michael Kodas).

Ewell, P.L. 1995. The Oakland-Berkeley Hills Fire of 1991. USDA Forest Service Gen. Tech. Rep. PSW-GTR-158.

Fagan, K. 2014. "Unprecedented" Winter Wildfires in Far Northern California. *San Francisco Chronicle*, January 6.

Farmer, J. 2013. *Trees in Paradise: A California History.* New York: Norton.

Federal Emergency Management Agency. 1992. The East Bay Hills Fire Oakland-Berkeley, California. U.S. Fire Administration/Technical Report Series USFA-TR-060. Washington, DC: FEMA.

———. 2002. Fires in the Wildland/Urban Interface. U.S. Fire Administration, Topical Fire Research Series 2(16).

Fire Adapted Communities Coalition. 2014. Guide to Fire Adapted Communities.

Flannigan, M.D., Krawchuk, M.A., de Groot, W.J., Wotton, B.M., and Gowman, L.M. 2009. Implications of Changing Climate for Global Wildland Fire. *International Journal of Wildland Fire* 18, 483–507.

Frammolino, R. 1991. Tragedy Renews Calls for Ban on Wooden Shingles: Prevention: Some communities require that roofing materials be fire-resistant, but measures that would have imposed statewide standards have not been able to get past opposition. *Los Angeles Times,* October 22, 1991.

Friederici, P. 2009. California Prepares for the Next Burn: Officials—and Homeowners—Start to Accept the Inevitability of Wildfire. *High Country News,* June 1, p. 20.

Gaillard, C., Glantz, M., Kelman, I., Wisner, B., Delica-Willison, Z., and Keim, M. 2014. Taking the "Naturalness" out of Natural Disaster (Again). *Natural Hazards Observer,* January, 1, 14–16.

Gale, W., Houser, S., and Scholz, J.K. 1996. Distributional Effects of Fundamental Tax Reform. In *Economic Effects of Fundamental Tax Reform,* ed. H. Aaron and W. Gale, ch. 8. Washington, DC: Brookings Institution.

Gamage, D. 2009. Coping through California's Budget Crises in Light of Proposition 13 and California's Fiscal Constitution, Proposition 13 at 30. In *After the Tax Revolt: Proposition 13 Turns 30,* ed. J. Citrin and I. Martin, ch. 4, 51–67. Berkeley, CA: Public Policy Press.

Gandy, M. 2002. *Concrete and Clay: Reworking Nature in New York City.* Cambridge, MA: MIT Press.

Gorte, R. 2013. The Rising Cost of Wildfire Protection. Bozeman, MT: Headwaters Economics. http://headwaterseconomics.org/wildfire/fire-costs-background/.

Green, M. 2013. The Flammable West: Mega-Fires in the Age of Climate Change (with real-time fire map). *KQED News,* August 9.

Greenwood, D., and Brown, T. 2003. *An Overview of Colorado's State and Local Tax Structures.* Colorado Springs: Center for Colorado Policy Studies.

Gude, P., Rasker, R., and van den Noort, J. 2008. Potential for Future Development on Fire-prone Lands. *Journal of Forestry* 106, 198–205.

Hammer, R.B., Radeloff, V.C., Freid, J.S., and Stewart, S.I. 2007. Wildland-Urban Interface Housing Growth during the 1990s in California, Oregon, and Washington. *International Journal of Wildland Fire* 16, 255–65.

Handy, R. 2014. Wildfires Fan Growth of Private Firefighting Crews. *The Coloradoan,* May 31. www.coloradoan.com/story/money/2014/06/01/wildfires-fan-growth-private-firefighting-crews/9744857/.

Handy, R.M. 2014. Waldo Canyon Fire: Two Years Later, a Neighborhood Is Reborn. *Colorado Springs Gazette,* June 23. http://gazette.com/waldo-canyon-fire-two-years-later-a-neighborhood-is-reborn/article/1521839#ChUhyFfEPhMb5V6L.99.

Harvey, D. 1973. *Social Justice and the City.* Baltimore: Johns Hopkins University Press.

———. 1996. *Justice, Nature and the Geography of Difference.* Oxford, UK: Blackwell.

Headwaters Economics. 2009. *Solutions to the Rising Costs of Fighting Fires in the Wildland-Urban Interface*, updated ver. December. Bozeman, MT: Headwaters Economics.

Hewitt, K. 1983. *Interpretations of Calamity*. London: Unwin Hyman.

———. 1997. *Regions of Risk: A Geographical Introduction to Disasters*. Essex, UK: Addison Wesley Longman.

Heynen, N., Kaika, M., and Swyngedouw, E. 2006. *In the Nature of Cities: Urban Political Ecology and the Politics of Urban Metabolism*. New York: Routledge.

Hobbs, R.J., et al. 2006. Novel Ecosystems: Theoretical and Management Aspects of the New Ecological World Order. *Global Ecology and Biogeography* 15:1, 1–7.

Hoffer, P. 2006. *Seven Fires: The Urban Infernos That Reshaped America*. New York: Public Affairs.

Holden, Z.A., Morgan, P., Crimmins, M.A., Steinhorst, R.K., and Smith, A.M. 2007. Fire Season Precipitation Variability Influences Fire Extent and Severity in a Large Southwestern Wilderness Area, United States. *Geophysical Research Letters* 34:16, L16708, doi:10.1029/2007GL030804.

Holthaus, E. 2016. We Need to Talk about Climate Change: Tragedies like the Fort McMurray Fire Make It More Important, Not Less. *Slate*, May 6.

Howard Jarvis Taxpayers Association. 2012. http://firetaxprotest.org/wp-content/uploads/2012/10/PR1003121.pdf (accessed August 22, 2015).

Hull, T. 1991. Oakland Fire Spurs Effort to Ban Untreated Wood Roofing in California. *San Francisco Examiner*, November 2.

Ingalsbee, T. 2010. *Getting Burned: A Taxpayer's Guide to Wildfire Suppression Costs*. Eugene, OR: Firefighters United for Safety, Ethics and Ecology.

International Association of Wildland Fire. 2013. www.iawfonline.org/pdf/WUI_Fact_Sheet_08012013.pdf (accessed May 25, 2015).

———. 2013. Yarnell Hill Fire Serious Accident Investigation Report. September 23.

Johnson, S. 2011. The Destruction Was Just the First Wave of Pain for Many Survivors of the Oakland Hills Fire. *Inside BayArea*, October 19. www.insidebayarea.com/oaklandtribune/localnews/ci_19148957.

Jolly, W.M., Cochrane, M.A., Freeborn, P.H., Holden, Z.A., Brown, T.J., Williamson, G.J., and Bowman, D.M. 2015. Climate-induced Variations in Global Wildfire Danger from 1979 to 2013. *Nature Communications* 6.

Kaika, M. 2005. *City of Flows: Modernity, Nature, and the City*. New York: Routledge.

Keil, R., and Desfor, G. 2004. *Nature and the City: Making Environmental Policy in Toronto and Los Angeles*. Tucson: University of Arizona Press.

Kemp, R.L. 1980. *Coping with Proposition 13.* Lexington, MA: Lexington Books.
Kolbert, E. 2016. Fort McMurray and the Fires of Climate Change. *New Yorker,* May 5.
Lave, R. 2012. Neoliberalism and the Production of Environmental Knowledge. *Environment and Society* 3, 19–38.
Laymance Real Estate Company. 1911. What the Development of the Foothills Has Meant to Oakland's Growth. In *Greater Oakland,* ed. E. Blake, 261–68. Oakland, CA: Pacific Publishing.
———. 1911. The Protection Rockridge Affords. Oakland, CA: Laymance Real Estate Company.
Liverman, D.M. 1990. Drought Impacts in Mexico: Climate, Agriculture, Technology, and Land Tenure in Sonora and Puebla. *Annals of the Association of American Geographers* 80:1, 49–72.
Lochner, T. 2014. Oakland-Berkeley Hills Groups at Odds over Eucalyptus Fire Hazard Abatement Tactics. *Contra Costa Times,* October 1.
LSA Associates Inc. 2009. Wildfire Hazard Reduction and Resource Management Plan for the East Bay Regional Park District. Draft Report, July 20.
Mai-Duc, C. 2016. Brown's Budget Earmarks Big Money for Natural Disasters. *Los Angeles Times,* January 8.
Marlon, J.R., Bartlein, P.J., Gavin, D.G., Long, C.J., Anderson, R.S., Briles, C.E., Brown, K.J., Colombarolig, D., Halletth, D.J., Poweri, M.J., Scharfj, E.A., and Walsh, M.K. 2012. Long-term Perspective on Wildfires in the Western USA. *Proceedings of the National Academy of Sciences* 109, E535–E543.
Marris, E. 2013. *Rambunctious Garden: Saving Nature in a Post-wild World.* New York: Bloomsbury USA.
Massey, N. 2012. Colorado Fire Follows in Pine Beetles' Tracks. *Scientific American, ClimateWire,* June 20.
McClintock, N. 2011. From Industrial Garden to Food Desert: Demarcated Devaluation in the Flatlands of Oakland, California. In *Cultivating Food Justice: Race, Class and Sustainability,* ed. A. Alkon and J. Agyeman, 89–120. Cambridge, MA: MIT Press.
Moritz, M.A., Batllori, E., Bradstock, R.A., Gill, A.M., Handmer, J., Hessburg, P.F., Leonard, J., McCaffrey, S., Odion, D.C., Schoennagel, T., and Syphard, A.D. 2014. Learning to Coexist with Wildfire. *Nature* 515, 58–66.
Muir, F. 1989. Wood-Roof Ban Sent to Bradley on 11–1 Vote: Industry in Near-Panic. *Los Angeles Times,* July 13.
Muller, E., and Dooling, S. 2011. Sustainability and Vulnerability: Integrating Equity into Plans for Central City Redevelopment. *Journal of Urbanism* 4:3, 201–22.
Mustafa, D. 1998. Structural Causes of Vulnerability to Flood Hazard in Pakistan. *Economic Geography* 74:3, 289–305.

———. 2005. The Production of Urban Hazardscape in Pakistan: Modernity, Vulnerability and the Range of Choice. *Annals of the Association of American Geographers* 95:3, 566–86.
National Fire Protection Association. 1992. The Oakland/Berkeley Hill Fire. National Wildland/Urban Interface Fire Protection Initiative.
———. 2013. Largest Loss Wildland Fires. NFPA Fire Incident Data Organization database. www.nfpa.org/research/reports-and-statistics/outdoor-fires/largest-loss-wildland-fires (accessed June 4, 2015).
———. 2015. Deadliest Incidents Resulting in the Deaths of 8 or More Firefighters. www.nfpa.org/research/reports-and-statistics/the-fire-service/fatalities-and-injuries/deadliest-incidents-resulting-in-the-deaths-of-8-or-more-firefighters (accessed July 13, 2015).
National Wildfire Coordinating Group. 2014. Wildland Urban Interface Wildfire Mitigation Desk Reference Guide. PMS 051. August.
Nowak, D.J. 1993. Historical Vegetation Change in Oakland and Its Implication for Urban Forest Management. *Journal of Arboriculture* 19:5.
Oakland Tribune. 1923. Oakland Hills' Forest Mantle All Handmade. July 15.
O'Brien, B. 2005. Ubiquitous Eucalyptus: How an Aussie Got Naturalized. *Bay Nature* (July–September), 1–6.
Oliver-Smith, A. 2002. Theorizing Disasters: Nature, Culture, Power. In *Culture and Catastrophe: The Anthropology of Disaster,* ed. S.M. Hoffman and A. Oliver-Smith. Santa Fe, NM: School of American Research Press.
Orenstein, N. 2015. Hills Group Sues FEMA over Plan to Cut Down Trees. *Berkeleyside,* March 23.
Orsi, J. 2004. *Hazardous Metropolis: Flooding and Urban Ecology in Los Angeles.* Berkeley: University of California Press.
Pelling, M. 2003. Toward a Political Ecology of Urban Environmental Risk. In *Political Ecology: An Integrative Approach to Geography and Environment-Development Studies,* ed. K.S. Zimmerer and T.J. Basset, 73–93. New York: Guildford.
Perkins, H. 2011. Gramsci in Green: Neoliberal Hegemony through Urban Forestry and the Potential for a Political Ecology of Praxis. *Geoforum* 42, 558–66.
Peterson, T.C., Heim Jr., R.R., Hirsch, R. Kaiser, D.P., Brooks, H., Diffenbaugh, N.S., Dole, R.M., Giovannettone, J.P., Guirguis, K., Karl, T.R., Katz, R.W., Kunkel, K., Lettenmair, D., McCabe, G.J., Paciorek, C.J., Ryberg, K.R., Schubert, S., Silva, V.B.S., Stewart, B.G., Vecchia, A.V., Villiarini, G., Vose, R.S., Walsh, J., Wehner, M., Wolock, D., Wolter, K., Woodhouse, G.A., and Wuebbles, D. 2013. Monitoring and Understanding Changes in Heat Waves, Cold Waves, Floods and Droughts in the United States: State of Knowledge. *Bulletin of the American Meteorological Society* 94:6, 821–34.
Pielke, R.A. 2002. Policy, Politics and Perspective. *Nature* 416:368.

Pincetl, S., Rundel, P.W., De Blasio, J.C., Silver, D., Scott, T., Keeley, J.E., and Halsey, R. 2008. It's the Land Use Not the Fuels: Fires and Land Development in Southern California. *Real Estate Review* 37, 25–42.

Pyne, S.J. 1982. *Fire in America: A Cultural History of Wildland and Rural Fire.* Princeton, NJ: Princeton University Press.

Ray-Bennett, N.S. 2007. Environmental Disasters and Disastrous Policies: An Overview from India. *Social Policy and Administration* 41:4, 419–24.

Robbins, P. 2007. *Lawn People: How Grasses, Weeds, and Chemicals Make Us Who We Are.* Philadelphia: Temple University Press.

Rodrigue, C.M. 1993. Home with a View: Chaparral Fire Hazard and the Social Geographies of Risk and Vulnerability. *California Geographer* 33, 105–18.

Romero-Lankao, P., Smith, J.B., Davidson, D.J., Diffenbaugh, N.S., Kinney, P.L., Kirshen, P., Kovacs, P., and Villers Ruiz, L. 2014. North America. In *Climate Change 2014: Impacts, Adaptation, and Vulnerability. Part B: Regional Aspects. Contribution of Working Group II to the Fifth Assessment Report of the Intergovernmental Panel on Climate Change,* ed. V.R. Barros, C.B. Field, D.J. Dokken, M.D. Mastrandrea, K.J. Mach, T.E. Bilir, M. Chatterjee, K.L. Ebi, Y.O. Estrada, R.C. Genova, B. Girma, E.S. Kissel, A.N. Levy, S. MacCracken, P.R. Mastrandrea, and L.L. White, 1439–98. Cambridge, UK/New York: Cambridge University Press.

Ross, J. 2009. Proposition 13 Thirty Years Later: What Has It Meant for Governance and Public Services? In *After the Tax Revolt: Proposition 13 Turns 30,* ed. J. Citrin and I. Martin, ch. 9, 135–40. Berkeley, CA: Public Policy Press.

Scawthorn, C., Eiginger, J.M., and Schiff, A.J., eds. 2005. *Fire Following Earthquake.* Reston, VA: American Society of Civil Engineers.

Self, R.O. 2003. *American Babylon: Race and the Struggle for Postwar Oakland.* Princeton, NJ: Princeton University Press.

Serna, J. 2014. California Drought Brings "Unprecedented" Fire Danger. *Los Angeles Times,* January 18.

Sibold, J.S., Veblen, T.V., and Gonzalez, M.E. 2006. Spatial and Temporal Variation in Historic Fire Regimes in Subalpine Forests across the Colorado Front Range in Rocky Mountain National Park, Colorado, USA. *Journal of Biogeography* 32, 631–47.

Silva, F., and Barbour, E. 1999. *The State-Local Fiscal Relationship in California: A Changing Balance of Power.* San Francisco: Public Policy Institute of California.

Simon, G.L. 2014. Vulnerability-in-Production: A Spatial History of Nature, Affluence, and Fire in Oakland, California. *Annals of the Association of American Geographers* 104:6, 1199–1221.

Simon, G.L., and Dooling, S. 2013. Flame and Fortune in California: The Material and Political Dimensions of Vulnerability. *Global Environmental Change* 23:6, 1410–23.

Simon, R. 1989. City, County Take Different Paths on Wood Roofs Issue. *Los Angeles Times,* July 7.

Slack, G. 2004. In the Shadow of Giants: The Redwoods of the East Bay Hills. *Bay Nature,* July 1. https://baynature.org/articles/in-the-shadow-of-giants/.

Somer, W.T., Coloff, S.G., and Conard, S.G. 2011. Fire Regimes. In *Synthesis of Knowledge: Fire History and Climate Change,* ch. 3. Project 09-02-1-09. Boise, ID: Joint Fire Science Program.

Sommers, W.T. 2008. The Emergence of the Wildland-Urban Interface Concept. *Forest History Today* (Fall), 12–18.

Sovern, D. 2011. Families Still Struggling with Loss 20 Years after Oakland Firestorm. *CBS SF Bay Area,* October 19. http://sanfrancisco.cbslocal.com/2011/10/19/families-still-struggling-with-loss-20-years-after-oakland-firestorm/.

State of California. 1991. The 1991 East Bay Firestorm. California Legislative State Senate. R. Torres, Chairman. Senate Committee on Toxics and Public Safety Management.

———. 2001. Oral History Interview with Tom Bates. California State Archives. State Government Oral History Program. By L. McGarrigle, Regional Oral History Office, University of California.

Stengers, I. 2011. "Another Science Is Possible!" A Plea for Slow Science. Inaugural lecture, Chair Willy Calewaert 2011–12. Faculté de Philosophie et Lettres, Université Libre de Bruxelles. December 12.

Stephens, S.L., Moghaddas, J.J., Ediminster, C., Fiedler, C.E., Hasse, S., Harrington, M., Keeley, J.E., McIver, J.D., Metlen, K., Skinner, C.N., and Youngblood, A. 2009. Fire Treatment Effects on Vegetation Structure, Fuels, and Potential Fire Severity in Western U.S. Forests. *Ecological Applications* 19, 305–20.

Stern, S. 1991. Firestorm Heats Up Debate over Development in Hills. *Oakland Tribune.*

Swyngedouw, E. 2004. *Social Power and the Urbanization of Water: Flows of Power.* Oxford, UK: Oxford University Press.

Taylor, T. 2013. UC Berkeley Expert Talks about Hillside Tree Removal Plan. *Berkeleyside,* June 11.

The Realty Syndicate. 1911. Realty Syndicate. In *Greater Oakland,* ed. E. Blake. Oakland, CA: Pacific Publishing.

Theobald, D.M., and Romme, W.H. 2007. Expansion of the US Wildland-Urban Interface. *Landscape and Urban Planning* 83, 340–54.

Troy, A., and Romm, J. 2006. An Assessment of the 1998 California Natural Hazard Disclosure Law (AB 1195). California Policy Research Center, University of California, Berkeley.

Tyrell, I. 1999. *True Garden of the Gods: California-Australian Environmental Reform 1860–1930.* Berkeley: University of California Press.

Union of Concerned Scientists. 2013. Is Global Warming Fueling Increased Wildfire Risks? www.ucsusa.org/global_warming/science_and_impacts/impacts/global-warming-and-wildfire.html#.VWyS5lwrjdk.

United States Census Bureau. 2016. www.census.gov/popclock/data_tables.php?component=growth (accessed April 19, 2016).

United States Department of Agriculture. 1995. Comparison of Fuel Load, Structural Characteristics and Infrastructure before and after the Oakland Hills "Tunnel Fire." USDA Forest Service Gen. Tech. Rep. PSW-GTR-158.

———. 2013. Wildfire, Wildlands, and People: Understanding and Preparing for Wildfire in the Wildland-Urban Interface. USDA Rocky Mountain Research Station, January.

U.S. Fire Administration; Federal Emergency Management Agency; Department of Homeland Security. 2002. Fires in the Wildland/Urban Interface. *Topical Fire Research Series* 2:16, 1–4. www.usfa.dhs.gov/downloads/pdf/tfrs/v2i16-508.pdf (accessed May 13, 2009).

U.S. Forest Service. n.d. National Forests on the Edge: Development Pressures on America's National Forests and Grasslands. www.fs.fed.us/openspace/fote/national_forests_on_the_edge.html (accessed May 25, 2015).

Walker, R. 1995. Landscape and City Life: Four Ecologies of Residence in the San Francisco Bay Area. *Ecumene* 2:1, 33–64.

———. 2004. Industry Builds Out the City: The Suburbanization of Manufacturing in the San Francisco Bay Area 1850–1940. In *The Manufactured Metropolis*, ed. Robert Lewis, 92–123. Philadelphia: Temple University Press.

Water Distribution Planning Division. 1992. Analysis of Broadway Terrace Pressure Zone B4Aa. East Bay Municipal Utility District, Oakland, CA.

Watts, M. 1983. On the Poverty of Theory: Natural Hazards Research. In *Interpretation of Calamity: From the Viewpoint of Human Ecology*, ed. Kenneth Hewitt, 231–62. Boston: Allen & Unwin.

———. 1983. *Silent Violence: Food, Famine, and Peasantry in Northern Nigeria*. Athens: University of Georgia Press.

Westerling, A.L., Hidalgo, H.G., Cayan, D.R., and Swetnam, T.W. 2006. Warming and Earlier Spring Increase Western US Forest Wildfire Activity. *Science* 313, 940–43.

Williams, R. 1975. *The Country and the City*. Oxford, UK: Oxford University Press.

Williamson, T.B., Colombo, S.J., Duinker, P.N., Gray, P.A., Hennessey, R.J., Houle, D., Johnston, M.H., Ogden, A.E., and Spittlehouse, D.L. 2009. *Climate Change and Canada's Forests: From Impacts to Adaptation*. Sustainable Forest Management Network and Natural Resources Canada, Canadian Forest Service, Northern Forestry Centre, Edmonton, AB.

Wisner, B. 1993. Disaster Vulnerability: Scale, Power, and Daily Life. *GeoJournal* 30:2, 127–40.

Wisner, B., Blaikie, P., Cannon, T., and Davis, I. 2004. *At Risk: Natural Hazards, People's Vulnerability, and Disasters*, 2nd ed. London: Routledge.

Wolch, J., Pincetl, S., and Pulido, L. 2001. Urban Nature and the Nature of Urbanism. In *From Chicago to L.A.: Making Sense of Urban Theory*, ed. M. Dear, 367–402. London: Sage.

Yoshino, K. 2007. Another Way the Rich Are Different: "Concierge-Level" Fire Protection. *Los Angeles Times*, October 26. www.latimes.com/business/la-fi-richfire26oct26-story.html#page = 1.

Index

Aamodt, Jim, 178
acacia, 62
activists, community/environmental, 104, 134, 137, 166
adaptation measures, 39–41, 49–50, 207; and climate change, 118–120, 122–23*fig.*, 124; and eucalyptus, 132–33, 221n3; and reducing demand for development, 204; "Refining and expanding community adaptation," 190–94, 192–93*table*, 198
Adger, Neil, 26, 70
aerial tanker assistance, 181, 187
aesthetic cohesion of rebuilt homes, 175
affluence, 18–19, 21, 25–26, 29–30, 32–34, 36–37, 188, 189*fig.*; and afforestation, 64, 66–69, 133; and Claremont Hotel and Resort, 57; and estate-based wealth protection, 32–33, 36, 81–82, 85–88, 86*map*, 92, 105, 167, 170; and eucalyptus, 133, 139; and extraction activities, 58–59; and the Incendiary, 110, 133; and mortgage interest deductions, 201; and postdisaster reconstruction efforts, 103–5, 155, 166–67, 176; and PRAs (private responsibility areas), 197; and private firefighting industry, 179, 183–84, 202–3; and race/class, 81–83, 85–88, 86*map;* and redistribution of wealth, 74, 81–82, 88, 143, 218n26; and reducing demand for development, 201–3, 206; and tax revenues, 72, 74, 81–83, 85–88, 86*map*, 218n26; and Tunnel Fire (1991), 69, 82, 90–92. *See also* AVI (affluence-vulnerability interface)
afforestation, 62–70, 63*figs.*, 65*figs.*, 109, 133–34, 137
African Americans, 46, 72, 87–88, 140–41. *See also* minorities
agriculture, 59, 110, 204, 220n6
agroforestry, 62, 64
AIG (American International Group), 177–78; Wildfire Protection Unit, 177–78
airflow, 15; convection columns, 112; outflow boundaries, 186; tornado-like vortices, 112; vertical convection columns, 57
Alagona, Peter, 139–140
Alameda County (Calif.), 72, 142
Alaska, 46, 117
Albany (Calif.), 193
Alberta (Canada) fire (2016), 19, 42–43
"American Babylon: Race and Struggle for Postwar Oakland" (Self), 32
American Southwest, 117
American West, 8, 16, 18–19, 21, 24*fig.*, 26–27, 33–37, 57–58, 207–8; and afforestation, 68, 70; and climate change, 48–51, 75, 119, 122–23*fig.*; and disaster capitalism,

239

American West *(continued)*
170; and eucalyptus, 132, 139, 221n3; and fire suppression, 196; as "Flammable West," 39, 43, 110; and the Incendiary, 19, 31, 37, 39–41, 43–51, 48*fig.*, 70, 110–11, 114, 119, 132, 188, 189*fig.*, 221n3; and mountain pine beetles, 116–18; as "new fire regime," 46; population growth in, 27, 43, 45–47, 119, 207; and postdisaster reconstruction efforts, 155, 158, 170; and private firefighting industry, 184; and race/class, 84; and reducing demand for development, 206; studies of large fires in, 44, 50; and tax revenues, 74, 77, 81, 84, 145–46; and variegated vulnerabilities, 93; and wood shingle industry, 37, 126. *See also names of Western cities, states, and regions*
Amito Pump, 159*maps*, 160
Angel Island, 137
Anglo-Americans, 134. *See also* whites
Annadel State Park (Sonoma County), 137
anorexia, 100
Anthropocene, 114, 220n6
antitax movement, 72, 77, 82, 84–85; and SRA fees, 145; and WPAD fees, 144–45. *See also* Proposition 13
anxieties: and the Incendiary, 112–13; and PRAs (private responsibility areas), 196; and Tunnel Fire (1991), 2, 90, 93, 95. *See also* fear
Arcata (Calif.), 49–50
architects/architecture, 170–71
aridity, 49, 75, 111, 117, 214n28, 220n14
Arizona: and the Incendiary, 46, 48*fig.*; Yarnell Hill Fire, 37, 185–88, 196
Armed Forces, U.S., 177
arsonists, 19–20, 31, 40–41
Asian Americans, 46, 72, 141. *See also* minorities
austerity, 177
Australasia, 50–51
Australia: "Prepare, stay and defend, or leave early" policy, 195, 195*fig.*
AVI (affluence-vulnerability interface), 18, 20, 22, 25, 29–32, 38, 207–8; and climate change, 120–21; and the Incendiary, 40, 47–48, 81, 109, 120–21, 132, 190; and reducing demand for development, 198*fig.*; and Tunnel Fire (1991), 70

Baker, John, 154
barbeques, 193

baselines, 137, 170
Bates, Tom, 124–25
bathrooms, number of, 174, 176
bay laurel (*Umbellularia californica*), 131, 137, 138*fig.*
Beck, Ulrich, 112
bedrooms, number of, 171, 173*fig.*, 176
Beetle Incident Management Organization, 118
Bel Air fire (1961), 127
belonging, 139–142, 175
benchmarking, historical, 137
Berkeley (Calif.): and afforestation, 35; City Council, 126, 137; and Claremont Hotel and Resort, 55, 57; and eucalyptus, 131–34, 137; and HEF (Hills Emergency Forum), 193; High School, 97; and Tunnel Fire (1991), 2, 11, 12*map*, 13*map*, 14, 16, 55, 57, 124; UC Berkeley, 2, 57, 125, 131, 134, 193, 221n1; and wood shingle industry, 124, 126
biodiversity, 50–51
biomass, 69, 134–35. *See also* fuel loads
Black Forest Fire (Colo., 2013), 174
Blackwater, 177
BLM (Bureau of Land Management), 200–201
blue stain fungus, 117
boreal forests, 42
Boulder County (Colo.), 204
Boulder Springs Ranch (Ariz.), 186
Braun, Bruce, 26–27
Brechin, Gary, 27, 32
Broadway Terrace, 154
Brown, Jerry, 121, 200
brownfield redevelopments, 198*fig.*, 206
brushfires, 11, 76

Cadillac Desert: The American West and Its Disappearing Water (Reisner), 27
Cal Fire, 75, 80–81, 121, 142, 144, 193, 220n14; Fire Hazard Severity Zone (FHSZ), 199
California, 8, 14, 16, 27, 37; attorney general, 127; budget of, 121, 142, 200; and climate change, 121; Department of Forestry, 124; and "drought package," 121, 200; FAIR Plan, 201–2; and fire codes, 127; "Home Rule Power," 74; and the Incendiary, 44, 47–49, 48*fig.*, 121; legislative ban on wood shingles, 124–28; and native vs. nonnative species, 137; Office of Emergency Services, 131; and Proposition 13, 32–33,

INDEX   241

72–75, 73*map*, 82–85, 88, 218n18, 218n26; and public disaster assistance, 83–84; "Separation of Sources Act," 74; State Forestry Board, 64; State Responsibility Areas (SRAs), 142–43, 145–46; and tax revenues, 32–33, 71–88, 73*map*. *See also names of cities and regions of California*

California Going, Going (Wood), 27

campfires, 112, 193

Canada: consulate, 125, 127; wood shingle industry, 125–28

canyons: and eucalyptus, 68; and road construction, 164; and Tunnel Fire (1991), 1, 15, 58, 151; and Yarnell Hill Fire (Arizona), 187

capitalism, 19, 27, 87, 183; disaster capitalism, 45, 170, 176; and risk society, 112

carbon sequestration, 205

cars, parked/abandoned, 7*fig.*, 103, 145, 150, 152, 154–55, 161–64, 163*fig.*; and rolling-car phenomenon, 162

cats, 2, 4–6

cattle, 59–60, 68

CCC (Claremont Conservation Conservancy), 131–32, 134–35, 137

Cedar Fire (San Diego County, Calif., 2003), 140*fig.*, 209nn3–4

Cedar Shake and Shingle Bureau (CSSB), 126–28; and law suit, 127

cedar shakes/shingles, 34–35, 124–29

Central Valley (Calif.), 15, 60

Cerro Grande Fire (Los Alamos, N.M.), 209nn3–4

chaparral, 43, 46, 136, 185

Chicago Fire (1871), 14, 155, 209n2

child care, 84

chimneys, 6, 7*fig.*, 93, 170

chronic fatigue syndrome, 98–99

Chubb Group of Insurance Companies, 178

cigarettes, discarded, 112

city planners, 68, 78, 91, 102–5, 112, 166, 188

Claremont Canyon, 137

Claremont Hotel and Resort, 55–58, 56*figs.*, 62, 134, 215n1

class, 81–88, 86*map*; class-based planning decisions, 212n23; elites, 32, 67–68, 183, 202–3; lower class, 33, 105, 207; middle class, 105, 198*fig.*, 207; and postdisaster reconstruction efforts, 171; and PRAs (private responsibility areas), 197; and reducing demand for development, 202–3,

205–7; and Tunnel Fire (1991), 96, 99, 105; working class, 72, 82, 99

classification of fire, 34, 111–13. *See also* firestorms

climate change, 34–35, 39, 45, 48–51, 75, 129, 214n28; and eucalyptus, 136; and firestorm as term, 111–12; and Fort McMurray fire (2016), 42–43; and the Incendiary, 111–12, 116, 118–124, 122–23*fig.*, 136; and mountain pine beetles, 116, 120, 124

climatic events, 15, 42, 214n28. *See also* climate change

coastal scrub, 15, 135; coastal scrub mosaic, 135; northern coastal scrub, 135

coast live oak (*Quercus agrifolia*), 5, 59, 135

coast redwood (*Sequoia sempervirens*), 58. *See also* redwood forest

Cold Fire, 180

collaborative efforts, 102–3, 158, 193–94

collectivized vulnerability reduction, 36, 91, 101–6, 166–67; collective good, 184; and individual actions, 105–6

Collins, Tim, 197

Colorado, 37; and home augmentation, 171, 174–76; and the Incendiary, 44–46, 48*fig.*, 49; largest/most destructive fires in, 44; and private firefighting industry, 178. *See also names of cities and regions in Colorado*

Colorado Springs (Colo.), 171, 174–76

combustion boxes, homes as, 69–70, 114–16, 115*fig.*. *See also* Duraflame firelogs

community fee programs, 198*fig.*, 202; and SRA fees, 142–43, 145–46; and WPAD fees, 143–46

commuters, 175, 203

Concrete and Clay: Reworking Nature in New York City (Gandy), 29

condominiums. *See* high-density residential housing

Congress, U.S., 22

conservation easements, 198*fig.*, 203–5

conservative homeowner politics, 32, 72, 109

Consumer Fire Products, 183

contaminated areas, 207

Contra Costa County (Calif.), 11, 142, 151

coping capacity, 95

coral reef damage, 50–51

CORE (Communities of Oakland Respond to Emergencies), 191

cypress, 62

Dansman, Raymond, 27
Davis, Mike, 29, 33, 46, 91, 171, 195–96, 212n23; "case for letting Malibu burn," 33, 91, 195–96; "upward social succession," 171
deaths in wildfires, 83, 207; of firefighters/police, 154, 164, 187–88, 196; and Tunnel Fire (1991), 8, 11, 93–95, 97, 103, 154, 161, 164–65, 196; and Yarnell Hill Fire (Arizona), 187–88, 196
decision-makers, 25, 29, 35, 70, 104–5, 212n23; and imposed risk, 91, 112; and the Incendiary, 119, 221n3
defensible space ordinances, 142, 191, 192–93*table*, 201
deforestation, 70
Denver (Colo.), 45
Department of Agriculture, U.S., 22
depoliticization, 20, 36, 41–42, 47–48, 58, 110–11, 121, 132–34
depression, 95
The Destruction of California (Dansman), 27
Deukmejian, George, 124
development, urban, 8, 18–23, 25–29, 37–38, 207–8; and adaptation measures, 80–81, 127, 190–94, 192–93*table;* and afforestation, 62–70, 63*figs.*, 65*figs.;* brownfield developments, 198*fig.;* and climate change, 120–21, 122–23*fig.;* deflecting development, 198*fig.*, 199, 205–8; disabling development, 198*fig.*, 199, 203–5, 208; discouraging development, 198–203, 198*fig.*, 208; and eucalyptus, 132–33, 139; and Fort McMurray fire (2016), 42–43; and the Incendiary, 45–48, 48*fig.*, 109–10, 114, 116, 120–21, 132–33, 188; mixed-use developments, 198*fig.*, 206; and PRAs (private responsibility areas), 195–98, 195*fig.;* and private firefighting industry, 184; and race/class, 87–88; "Reducing demand for development," 198–207, 198*fig.;* strategies to combat root causes of, 189–207; and tax revenues, 32–33, 77–81, 78*table*, 87–88; and Tunnel Fire (1991), 14–16, 77–78; and Yarnell Hill Fire (Arizona), 187–88. *See also* home construction; suburban landscapes
Diablo winds, 15, 55, 191
disabled persons, 3–4, 28, 94–95, 98; disability from work, 98
"disaster apartheid," 183
disaster capitalism, 45, 170, 176
distraction, debates of, 20–21, 34–35, 120, 130, 136, 221n3

drones, 181
droughts, 34, 39, 43, 45, 48, 50, 64, 75; and climate change, 118–19, 121, 124; and "drought package," 121, 200; Drought Severity Index, 185; and Yarnell Hill Fire (Arizona), 185
Duraflame firelogs, 69–70, 113–16, 115*fig.*

East Bay hills, 2, 17*map*, 193; and afforestation, 64, 67–70; and eucalyptus, 131–32, 135–38, 140–41; and extraction activities, 59–60; and HEF (Hills Emergency Forum), 193–94; and the Incendiary, 109–10, 131–32; and tax revenues, 80. *See also names of cities in East Bay hills*
East Bay Hills Firestorm. *See* Tunnel Fire (1991)
East Bay Municipal Utility District (EBMUD), 157, 160, 193
East Bay Regional Park District (EBRPD), 69, 80–81, 135–36, 193, 221n1
East Bay Watershed Company, 64
economic benefits, 8, 18–19, 37; and "fight like hell," 105; and firestorm as term, 116; and power line reconstruction, 104; and tax revenues, 78, 80, 86*fig.*
economic development/growth, 27; and deforestation/afforestation, 31, 55, 59, 70; and the Incendiary, 110, 116. *See also* development, urban
economic recessions, 144
ecosystem services, 43, 48, 198*fig.*, 205; PES (payments for ecosystem services), 205
education, 84, 96–97, 105, 175; fire education, 155, 190, 193–95, 199–200; public education activities, 142–43
El Cerrito (Calif.), 193
elderly persons, 3–4, 94–95, 100, 175
El Niño, 42, 214n28; Southern Oscillation (ENSO) events, 214n28, 220n14
emergency radio communications, 36, 76–77, 89–90, 149–155, 160, 167; and radio frequency incompatibilities, 76, 150, 158
emergency shelter deployment, 187
Environmental Protection Agency (EPA), 180
estate-based wealth protection, 170; and postdisaster reconstruction efforts, 167; and tax revenues, 32, 81–82, 85–88, 86*map;* and Tunnel Fire (1991), 33, 36, 92, 105
ethical considerations, 207; and private firefighting industry, 178, 183–84

INDEX 243

eucalyptus (*Eucalyptus globulous*), 35, 62, 64–70, 65*figs.*, 131–142, 138*figs.*, 140*fig.*; blue gum eucalyptus, 64, 137; dead/damaged trees, 75; eradication of, 131–32, 135–37, 138*figs.*, 141; eucalyptus groves, 131, 134–36, 138*fig.*; and flammability, 132–36, 139, 141; Japanese eucalyptus, 65*fig.*; and owls, 139, 138*fig.*; and postdisaster reconstruction efforts, 169, 169*fig.*; and race/class, 139–42; scapegoating of, 135, 137; sign on, 169, 169*fig.*; thinning of, 131, 137, 138*figs.*; and Tunnel Fire (1991), 1–3, 132, 134–37, 141; volatile oils of, 134; vs. native species, 134, 136–38, 140–42

evacuations: and blocked routes, 103, 152–55, 165; and fire preparedness measures, 191; and postdisaster reconstruction efforts, 160–65, 162*fig.*, 163*fig.*; and private firefighting industry, 181–82, 184; and Tunnel Fire (1991), 3–6, 77, 80, 94–96, 103, 149, 151–55; and Yarnell Hill Fire (Arizona), 185

evapotranspiration rates, 49

exotic species, 137

extraction activities, 18–19, 31–32, 58–60, 60*fig.*, 88; and access roads, 55, 57, 60–62, 61*map*; and Fort McMurray fire (2016), 42–43; and the Incendiary, 109; mineral extraction, 59; oil extraction, 19, 42–43. *See also* timber extraction

exurban landscapes, 8, 15, 22, 25, 31; and the Incendiary, 47–48, 114

FAIR Plan (Calif.), 201–2

Farm Bill (2014), 118

fears, 207; and eucalyptus, 137–38; and the Incendiary, 112–14, 115*fig.*; and postdisaster reconstruction efforts, 163; and PRAs (private responsibility areas), 197; and Tunnel Fire (1991), 2, 90, 95. *See also* anxieties

federal mitigation funds, 188; and eucalyptus, 131–34; and mountain pine beetles, 118, 120; and reducing demand for development, 198*fig.*, 200–202, 204; and tax revenues, 83–84, 175; and Yarnell Hill Fire (Arizona), 187

FEMA (Federal Emergency Management Agency), 68–69, 75; and eucalyptus, 131–35, 137, 221n1; and power lines, 164; and reducing demand for development, 200–201; unified methodology endorsed by, 131, 137

"Fighting fire with fire," 195–98, 195*fig.*, 201

"Fire Adapted Communities," 23

Fire Administration, U.S., 207

fire behavior, 16, 186, 196

firebreaks, 134, 142; Highway 13 as, 156; Highway 24 as, 153

Firebreak Spray Systems (FSS), 177–78

fire chiefs, 193

fire codes, 41, 127, 190–94, 192–93*table*, 199, 204

fire departments, 50; and budgets, 32, 74–77, 79–84, 119, 121, 142, 145, 218n18; and emergency radio communications, 36, 76–77, 89–90, 149–155, 160, 167; fire alarm declarations, 150; at Fourth of July parade, 96; ICS (Incident Command System), 76, 89, 150–54; and mutual aid, 76–77, 80–81, 151, 157–58, 181–82; new fire stations, 78–79, 81; and nozzle hookup incompatibilities, 76, 157–58; and postdisaster reconstruction efforts, 155–57; and radio frequency incompatibilities, 76, 150, 158; and red-flag fire days, 75, 191; and tax revenues, 32–33, 74–84, 218n18; and Tunnel Fire (1991), 75–76, 89–90, 98, 103, 149–155; and WPAD fees, 143. *See also* firefighting/firefighters

fire disasters, 113–14, 116, 121. *See also* firestorms

"fire exclusion paradigm," 23

Firefighters United for Safety, Ethics, and Ecology, 183

firefighting/firefighters, 14; and aerial tanker assistance, 181, 187; and Claremont Hotel and Resort, 57; and the Incendiary, 45–47, 119, 188–89, 189*fig.*, 220n9; loss of life in fire, 154, 187–88; and PRAs (private responsibility areas), 195–98, 195*fig.*; private firefighting industry, 18, 35, 37, 45, 177–184, 194, 202–3; and reducing demand for development, 200, 202; sheltering in concrete block garage/swimming pool, 152–53; and Tunnel Fire (1991), 11, 14, 36, 55, 57, 75–77, 89–90, 98, 103, 149–155; and Yarnell Hill Fire (Arizona), 186–88. *See also* fire departments

FireIce, 180

fire ladders, 131, 134, 192–93*table*

fire managers, 49–50

fire mapping, 142, 198*fig.*, 199–200; Cal Fire's Fire Hazard Severity Zone (FHSZ), 199; fire zone severity mapping projects, 45

fire marshals, 50, 126, 144, 200

fire prevention, 80–81, 127–28, 190–94, 192–93*table;* and private firefighting industry, 177; and SRA Fire Prevention Benefit fee program, 142–43, 145–46; and Wildfire Prevention Assessment District (WPAD) fee, 143–46
fire regimes, 16, 220n7; frequency of, 16, 17*map;* larger, more intense fires, 44, 50, 118, 121, 122–23*fig.*, 196–97; longer in duration, 49–50, 75, 119, 122–23*fig.;* "new fire regime," 46; periodic wildfires, 16, 17*map*, 23, 43, 49, 114, 116, 196
fire retardants: as defense strategy, 179–180; fire-retardant/fire-resistant shingles, 124–28, 221n24; foam application products, 178–180, 182–83; gel application products, 179–180, 182–83; like "snake oil," 179–180; and private firefighting industry, 178–183; "water on steroids," 180
fire shelters, 187
firestorms, 110–16; and Claremont Hotel and Resort, 55–57, 56*fig.;* defined, 112–13; as fire disasters, 113–14, 116, 121; firestorm as term, 34, 110–13, 115*fig.*, 116, 129; and homes like Duraflame firelogs, 113–16, 115*fig.;* military context of, 111; and private firefighting industry, 179; root causes of, 38. *See also names of firestorms*
first responders, 128, 149, 191, 195. *See also* firefighting/firefighters; police
flammability, 8, 19–20; and afforestation, 35, 69, 133; and climate change, 119–121; and density of flammable structures, 48, 132–33, 140*fig.;* and eucalyptus, 132–36, 139, 141; flammable landscapes, 19, 31–32, 40–42, 45; homes as combustion boxes, 69–70, 114–16, 115*fig.;* and the Incendiary, 39–42, 45, 47, 109–10, 114–16, 115*fig.*, 220n14; and mountain pine beetles, 118; seeming inevitability of, 39–40; and tax revenues, 70–71, 75, 78; and wood shingle industry, 126; and Yarnell Hill Fire (Arizona), 186
"The Flammable West: Mega Fires in the Age of Climate Change" (article), 39, 43, 110
flatlands, 21, 33, 82–88, 86*map*, 93
floods, 50, 207, 220n14
flying brands, 23, 69, 125, 134, 153, 165, 178–79, 191, 192–93*table*
foam application products, 178–183; compressed air foam systems, 180; health/environmental effects of, 180; as home kits, 180, 183; and PFOS (perfluorooctane sulfonates), 180–81; prefire application of, 182–83; and safety concerns, 182–83; telomer-based foams, 180
foothills, 59, 71, 73*map*, 158
forested landscapes, 15, 23, 32, 48–49, 58, 68–69; density of, 48, 58, 69, 80; fires in, 125, 165; and mountain pine beetles, 34, 116–18, 120, 124; and tree mortality, 117–18. *See also names of tree species*
Forest Products Association, 126
forestry activities, 62, 64, 68. *See also* lumber industry; timber extraction
Forest Service, U.S., 23, 47, 117; "Fire Adapted Communities," 23; "fire exclusion paradigm," 23; and reducing demand for development, 200–201
Fort McMurray fire (2016, Alberta, Canada), 19, 42–43
foundational characteristics, 19–20, 41, 132, 160
Fourth of July, 96
fraud, 96
free markets, 20, 59, 74, 170, 177
Fremont (Calif.), 193
Front Range (Colo.), 49
fuel loads, 15, 22, 37, 43–45, 69, 81–82; abandoned cars contributing to, 161–62; and Claremont Hotel and Resort, 57; and eucalyptus, 131, 134–36; and firestorm as term, 112; and home augmentation, 176; homes as combustion boxes, 69; and mountain pine beetles, 118; and SRA fees, 142; vegetative fuels, 22

Gandy, Matt, 29
gas-powered equipment, 192–93*table*, 193
gel application products, 179–180, 182–83; health/environmental effects of, 180; as home kits, 180, 183; prefire application of, 182–83; and safety concerns, 182–83
gender: gender-based violence, 92; and reducing demand for development, 205; and single women, 96, 102; and Tunnel Fire (1991), 92, 94, 96; and variabilities of cause, 94
gentrification, 206
geography, 21–23, 26–27, 33; and Claremont Hotel and Resort, 57; and the Incendiary, 46–47; multigeographical approach, 18; and race/class, 87; and tax revenues, 87; and Tunnel Fire (1991), 14–16, 18, 57, 91, 94; and variabilities of cause, 94; and WUI, 22–23

ghost towns, 196
goats, 143
Gold Rush, 58–59, 187
government policies, 19–20, 22, 32, 47; and SRA fees, 142–43, 145–46; urban policies, 74; and WPAD fees, 143–46. *See also* federal mitigation funds; state mitigation funds
Granite Mountain Hotshots, 186–88; loss of life in fire, 187–88
Grassetti, Dan, 131
grasslands, 15, 131, 165, 174; grasslands mosaic, 135; and Yarnell Hill Fire (Arizona), 185
grazing, 59–60, 68, 143
greenhouse gas emissions, 119
Grizzly Peak, 150–51
ground fires, 15
Grubensky, John, 154
Gwin Tank, 89–90, 159*maps*

Hahn, Kenneth, 124, 127
Hancock, Loni, 132
Harris, Elihu, 132
Harris Fire (San Diego County, Calif., 2007), 209n3
Harvey, David, 27, 87
Havens, Frank, 64, 66
hay production, 59
Hayward (Calif.), 193
Hazardous Metropolis: Flooding and Urban Ecology in Los Angeles (Orsi), 29
hazards, fire, 26, 28, 30, 37, 188, 193–94; and climate change, 120; and eucalyptus, 131, 135; fire hazard mapping, 142, 198*fig.*, 199–200, 202; hazardscapes, 26; and postdisaster reconstruction efforts, 162, 167; and private firefighting industry, 179–180; and race/class, 82, 84; and reducing demand for development, 198*fig.*, 199–200, 202, 207; and tax revenues, 77–78, 82, 84; and Tunnel Fire (1991), 91, 94, 101, 126, 135; and wood shingle industry, 126
HCN (Hills Conservation Network), 131, 135, 137–38; website of, 139
HEF (Hills Emergency Forum), 193–94; Staff Liaison Committee, 193
the "Heights". *See* Miller homestead
helicopters, 154
herbicides, 136
heroism, 154–55, 188
Hetch Hetchy Valley (Calif.), 60
Hewitt, Kenneth, 26

high-density residential housing, 13*map*, 15, 77–79, 79*fig.*
Highway 13, 156
Highway 24, 13*map*, 149, 153
Hiller Highlands, 151, 159*maps*, 160; Hiller Highlands Complex, 13*map*, 77
Hills Emergency Forum, 80–81
hillside landscapes, 21; and afforestation, 62–70, 63*fig.*, 65*fig.*; and eucalyptus, 131–34, 141; and high-density residential housing, 77–78; and hill area disaster plan, 150; and postdisaster reconstruction efforts, 104, 150, 155, 158, 160, 162, 164–65, 170–71, 175–76; and race/class, 82–83, 85–88, 86*map*; and reducing demand for development, 200; and tax revenues, 73*map*, 77–80, 82–83, 85–88, 86*map*, 175–76; and Tunnel Fire (1991), 1–4, 11, 14–15, 55, 57–58, 92, 101, 104, 150–51, 154; and WPAD fees, 144; and Yarnell Hill Fire (Arizona), 185–86. *See also* East Bay hills; Oakland Hills
Hiroshima, 93, 111
Hispanics, 46, 72, 141. *See also* minorities
history of wildfires, 16, 19, 21, 25–26, 29–31, 212n23; and afforestation, 68–70; and Claremont Hotel and Resort, 57; and fire mapping, 199; and firestorm as term, 34, 110–13; and fire suppression, 196; and home footprints, 171; and the Incendiary, 41, 46, 49, 109, 114, 120–21, 220n7; and race/class, 83, 85–88, 86*map*; and tax revenues, 78, 81, 83, 85–88, 86*map*; and timber extraction, 58; and Tunnel Fire (1991), 16, 31–32, 57, 78, 153; and WPAD fees, 144
HJTA (Howard Jarvis Tax Association), 143
Holocene, 220n6
home construction, 31–32, 34–35, 37, 45, 125, 207; and adaptation measures, 194; and aesthetic cohesion, 175; affordable housing options, 198*fig.*, 206–7; and afforestation, 62–70, 63*figs.*, 65*figs.*, 133–34; and bathrooms, number of, 174, 176; and bedrooms, number of, 171–74, 172–73*figs.*, 176; and Claremont Hotel and Resort, 55, 57; and climate change, 120–24, 122–23*fig.*; and defensible space ordinances, 142, 191, 192–93*table*, 201, 204; and density of flammable structures, 48, 132–33, 140*fig.*; and elevated garages, 162; and eucalyptus, 132–34, 140*fig.*; and fire codes, 41, 127, 190–94, 192–93*table*, 199,

246   INDEX

home construction *(continued)* 204; and fire prevention ordinances, 80–81, 127–28; and fire-safe standards, 176, 199, 201; and home augmentation, 37, 170–76, 172–73*figs.;* and home footprints, 171–74, 172–73*figs.*, 201; and homes like Duraflame firelogs, 69–70, 113–16, 115*fig.;* and the Incendiary, 109, 111, 113–16, 115*fig.*, 120–21, 129, 133, 146, 160; and indoor sprinkler requirements, 80–81; and modified eaves, 81, 125–26; and mortgage interest deductions, 198*fig.*, 201; and permitting procedures, 174–75; and postdisaster reconstruction efforts, 161–62, 170–76, 172–73*figs.;* and PRAs (private responsibility areas), 195–98, 195*fig.;* and race/class, 82; and reducing demand for development, 199–200, 203–7; and second homes, 201; and setback rules, 80; and tax revenues, 71, 77–82, 79*fig.;* and timber extraction, 58–60, 61*map;* and Waldo Canyon Fire (Colo., 2012), 171, 174; and wood shingle industry, 34–35, 114, 124–29; and Yarnell Hill Fire (Arizona), 187–88. *See also* development, urban; suburban landscapes
home densities, 16, 22–23
homeowners. *See* private property owners/rights
"Home Rule Power" (Calif.), 74
homes/structures damaged/destroyed by fire, 28, 44; and Claremont Hotel and Resort, 57; and eucalyptus, 140*fig.;* and firestorm as term, 112–13; and postdisaster reconstruction efforts, 156–57, 161–62, 166, 169; and private firefighting industry, 179; and tax revenues, 83; and Tunnel Fire (1991), 5–6, 8, 11, 14, 16, 57, 79*fig.*, 92–101, 103, 153–54; and Waldo Canyon Fire (Colo., 2012), 174. *See also* postdisaster reconstruction efforts
home survival, 98–99
household-scale analysis, 91–93, 101
house painters, 99
Humboldt County (Calif.), 75
hydrocarbons, 114

ICS (Incident Command System), 76, 89, 150–54
Idaho, 46, 48*fig.*
immigrants, 84
Imperial San Francisco: Urban Power, Earthly Ruin (Brechin), 27, 32
incendiaries, 40–41
the Incendiary, 18–20, 34–36, 38, 40–51, 70, 188–89, 189*fig.*, 207–8, 219n1, 221n24; as arsonist, 19–20, 31, 40–41, 194; character profile of, 43–51; combating the Incendiary: adaptation measures, 190–94, 192–93*table*, 198; combating the Incendiary: PRAs (private responsibility areas), 195–98, 195*fig.*, 201; combating the Incendiary: "Reducing demand for development," 198–207, 198*fig.;* as concealed/less visible, 34–35, 109–16, 119–121, 129–133, 146; depoliticization of, 20, 41–42, 47–48, 58, 110–11, 121, 132–34; foundational characteristics of, 19–20, 41, 132, 160; naturalization of, 34–35, 41, 47–48, 111–12, 115*fig.*, 116, 129, 132–33; as patient, 31, 38, 41, 190; repoliticization of, 20, 36, 47–48, 130, 132, 139; and SRA fees, 146; and tax revenues, 81; and WPAD fees, 146
individual freedoms/rights, 20–22; and "fight like hell," 105–6; and private firefighting industry, 181–84; and tax revenues, 32, 81; and Tunnel Fire (1991), 21–22, 105–6
Industrial Revolution, 220n6
in-fill practices, 46, 198, 198*fig.*, 205–7
inflation rates, 72
infrastructure, 58–60, 66, 68–69, 194; and postdisaster reconstruction efforts, 150; and reducing demand for development, 200; and SRA fees, 142; and tax revenues, 72, 80. *See also* power lines; roads; water infrastructure
insurance industry, 25, 28, 45; canceling plans in high-risk areas, 200; and comprehensive fire insurance plans, 83–85, 92, 178, 184, 197; higher insurance premiums, 199–202; and loss indemnification, 83, 188, 197; negotiations with, 96, 99, 102–3, 171; and postdisaster reconstruction efforts, 102–3, 105, 171, 176; and PRAs (private responsibility areas), 197; and private firefighting industry, 177–78, 181–84; and race/class, 83–85, 96, 105; and reducing demand for development, 198*fig.*, 199–202; rental insurance plans, 84; and settlement costs, 178; and single women, 96; and tax revenues, 83–85; and Tunnel Fire (1991), 33, 83, 92, 95–96, 99, 105; and United Policyholders (insurance holder advocacy program), 102–3
invasive species removal, 136, 141

INDEX 247

IPCC (Intergovernmental Panel on Climate Change), Fifth Report, 49–50; confidence levels in, 50

Kaufman, Jon, 132
Key Route mass transit rail system, 57, 64
Klein, Naomi, 170, 176, 183

land classifications, 23, 25; nonurban/urban land classifications, 23
land conversion, 47, 48*fig.*, 136
land cover change: and afforestation, 69; and Claremont Hotel and Resort, 56*figs.*, 57; and Tunnel Fire (1991), 56*figs.*, 57
land managers, 80
landslides, 207
land swaps, 204
land use planning/practices, 8, 14, 16, 22–23, 25–26, 29, 87; and agriculture, 59, 110; and the Incendiary, 109; and reducing demand for development, 199–201
Lawrence Berkeley National Laboratory, 193
Laymance Real Estate Company, 87–88, 140–41
Leland, Abigail, 62
lightning, 112
litter, ground, 75; and eucalyptus, 131, 134–36; removal of, 131, 135; in Tunnel Fire (1991), 1–2, 15, 69, 135–36
Liverman, Diana, 26
lodgepole pine, 117
logging activities/roads, 55, 57–62, 60*fig.*, 61*map*, 63*fig.*, 65*fig.*, 66, 215n2, 216nn7–8; and the Incendiary, 109
Los Alamos (N.M.), 209n3
Los Angeles (Calif.), 29, 212n23; and fire prevention ordinances, 127–28; and law suit, 127; Los Angeles Fire Department (LAFD), 127; and private firefighting industry, 182; and wood shingle industry, 124, 127–28
losses, 190; and climate change, 50; cost estimates of, 45; loss of life, 8, 11, 83, 93–95, 97, 103, 154, 161, 164–65, 187–88, 196, 207; of material items, 83, 92–94; and race/class, 83–85; and Tunnel Fire (1991), 8, 11, 14, 34, 83, 90, 92–95, 103, 106, 154, 161, 164–65, 196; and Yarnell Hill Fire (Arizona), 37, 187–88, 196
lucrative landscapes, 18–20, 31–32, 208; and eucalyptus, 133; and Fort McMurray fire (2016), 42–43; and the Incendiary, 45, 109–10, 112, 120, 133; and race/class, 85,

88; and reducing demand for development, 202; and tax revenues, 78, 85, 88
lumber industry, 58–61, 60*fig.*, 61*map*; and eucalyptus, 64

Mahogany Eucalyptus and Land Company, 66
Malibu (Calif.), 178; "case for letting Malibu burn," 33, 91, 195–96
management, 8, 15–16, 19–20, 22–23, 25, 29, 31, 208; and climate change, 119; and eucalyptus, 35, 67, 131–33, 139, 141, 221n3; forest management practices, 23, 42, 117–18; and HEF (Hills Emergency Forum), 193–94; and the Incendiary, 20, 34–36, 40, 109, 112–13, 116, 119, 131–33, 221n3; and mountain pine beetles, 117–18; and postdisaster reconstruction efforts, 149–150, 167; and reducing demand for development, 205; and SRA fees, 143
Manila, Battle of, 93
marginalization, 93, 96, 101, 197, 205–7
Marin County (Calif.), 49–50
marketing strategies: and Claremont Hotel and Resort, 57, 62; and eucalyptus, 62, 64; and private firefighting industry, 179–180, 184
material items, loss of, 83, 92–94, 99–101; books being written, 100–101; diaries, 100; mementos/keepsakes, 99–101; photographs, 99–100; symbolic/nostalgic value of, 99–100
media: and climate change, 42–43, 121; and "drought package," 121; and "the Flammable West," 39; major media market, 101; and PRAs (private responsibility areas), 197; and Tunnel Fire (1991), 101
Mediterranean climate, 15
mental health issues, 95
Middle East, 170
Miller, Joaquin, 62, 63*fig.*
Miller homestead, 62, 63*fig.*, 134
Minnesota, 181
minorities, 33, 46, 81–88, 86*map*, 102, 197. *See also* class; race; *names of minority groups*
Mission San José, 58
Missoula County (Mont.), 203–4
mitigation efforts, 16, 26, 39–41, 193–94; and Claremont Hotel and Resort, 57; cost of, 8, 18–19, 35–37, 44–45, 51; do-it-yourself mitigation, 180, 183; and eucalyptus,

mitigation efforts *(continued)* 131–35; federal mitigation funds, 83–84, 118, 120, 131–34, 200–202, 204; and the Incendiary, 116, 120, 131–32, 188; and mountain pine beetles, 118; and postdisaster reconstruction efforts, 150, 155, 158, 160, 162; and PRAs (private responsibility areas), 195–98, 195*fig.*; and private firefighting industry, 181, 183; and race/class, 83–85; and reducing demand for development, 200–202, 204; and SRA fees, 143, 146; state mitigation funds, 44–45, 83, 121, 175, 188, 190, 193, 200–202; and tax revenues, 75, 77, 79–80, 83–85; and Tunnel Fire (1991), 151, 155; and wood shingle industry, 126, 128; and WPAD fees, 143–44, 146; and Yarnell Hill Fire (Arizona), 186
mixed-use developments, 198*fig.*, 206
mobility, 94
Mongolian descent, 87–88, 140–41
Montana, 48*fig.*, 178, 203–4; Working Forest Project, 203–4
Montaña de Oro State Park (San Luis Obispo County), 137
Monterey pine (*Pinus radiata*), 64, 67, 69, 135
mortgage interest deductions, 198*fig.*, 201; and second homes, 201
mountain pine beetle (*Dendroctonus ponderosae*), 34, 116–18, 120, 124, 129, 220n9
Mountain Shadows (Colorado Springs), 174
Mustafa, Danish, 26
mutual aid, 76–77, 80–81, 151, 157–58; and private firefighting industry, 181–82

Napa County (Calif.), 200, 204
National Fire Protection Association (NFPA), 125, 209n2
national forests/grasslands, 47, 196. *See also* public lands
National Wildfire Coordinating Group, 22
Native American art collection, 100
native species, 35, 59, 62–70, 63*fig.*, 65*fig.*, 114, 116; and race/class, 141; restoration efforts, 137–38; vs. eucalyptus, 134, 136–38, 138*figs.*, 141. *See also* nonnative species; *names of species*
natural disasters, 113, 171
Natural Hazards Center (2014 report), 120
natural resources, 16, 19–20, 26–27, 29, 31–32, 58–60. *See also* extraction activities

Nature Communications (2015 article), 119–120
Nature Conservancy, 204
Nature study, 140*fig.*
neoliberal urban policy: and mutual aid, 76–77, 158, 182; and private firefighting industry, 177, 182; and rapid science, 30
neopopulism, 32, 81
Nevada, 46, 48*fig.*
New Mexico, 48*fig.*
New Urbanism, 46
New York City, 29
nongovernmental organizations (NGOs), 204
nonnative species: eucalyptus as, 35, 64, 65*fig.*, 67, 131–32, 134, 136–37, 138*figs.*, 141; and Miller homestead, 63*fig.*; nonnative pine forest, 131; and race/class, 141. *See also* native species
nonprofits, 47, 126, 202, 204
North Coast (Calif.), 60
Northern California, 49–50; and private firefighting industry, 178, 181; and reducing demand for development, 200; and winter wildfires, 75
North Hills Community Association, 168–69
North Hills Phoenix Association, 168
The North Oakland Hills Area Specific Plan, 68, 78

Oakland (Calif.), 21, 32–33, 208; and afforestation, 35, 66–67; City Council, 137; and Claremont Hotel and Resort, 55, 57; CORE (Communities of Oakland Respond to Emergencies), 191; and disaster capitalism, 170; and eucalyptus, 132, 137, 140–41, 221n1, 221n3; and fire preparedness measures, 191; and fire prevention ordinances, 127–28; and HEF (Hills Emergency Forum), 193–94; historical planning records, 153; and the Incendiary, 109, 132, 221n3; and nozzle hook-up incompatibilities, 76, 157–58; and operating budgets, 155, 160, 163–64, 166; and population growth, 58; and postdisaster reconstruction efforts, 36, 103–4, 150, 155–57, 162*fig.*, 170, 175; and race/class, 81–88, 86*map*, 140–41; and road construction, 60, 61*map*, 66; and SRA fees, 142; and tax revenues, 32–33, 71–88, 73*map*, 78*table*, 86*map*, 160, 175–76; as terminus of transcontinental railroad, 58; and timber extraction, 58, 61*map;* and water/power line upgrades, 36; Wildfire

INDEX   249

Prevention Assessment District (WPAD) fee, 143–46. *See also* fire departments; Tunnel Fire (1991)
Oakland Estuary, 60
Oakland Hills, 16, 31–33; and afforestation, 62–71, 63*fig.*, 65*fig.*; and assumed risk, 91–93; and eucalyptus, 131–142; and extraction activities, 58–60, 60*fig.*; and fire preparedness measures, 191; and the Incendiary, 111, 114, 116; "part of the city below, yet apart from it," 87; and race/class, 82–83, 85–88, 86*map*; and tax revenues, 75, 77–79, 82–83, 85–88, 86*map*. *See also* Tunnel Fire (1991)
Oakland Hills fire (September 22, 1970), 165–66
Oakland Hills Firestorm (1991). *See* Tunnel Fire (1991)
Oakland Tech, 97
oak trees, 131, 136–37
O'Connell, Jack, 126
oil extraction, 19, 42–43; oil as replacement of wood fuel, 64
Old Fire (San Bernardino County, Calif., 2003), 209nn3–4
open spaces, 14–15, 23, 79*fig.*, 110, 142, 156, 203–4
opportunism, financial, 169–170, 208
Orange County (Calif.), 182
Oregon, 47, 48*fig.*
origin of wildfires: and firestorm as term, 112; of Tunnel Fire (1991), 11, 13*map*, 79*fig.*, 89–90, 137, 159*maps*; of Yarnell Hill Fire (Arizona), 185
Orinda/Moraga (Calif.), 193
ornamental species, 63*fig.*, 65*fig.*, 68. *See also* nonnative species
Orsi, Jared, 29
owls, 139

Pacific Coast Ranges, 70
Pacific Gas & Electric (PG&E), 104, 152
Pacific Northwest, 128
Palo Seco Creek, 60*fig.*
parks. *See* public lands
Parkwoods Apartments (Oakland), 13*map*, 77
patriotism, American, 177
Peeples Valley (Ariz.), 185, 187
periodic wildfires, 16, 17*map*, 23, 43, 49, 114, 116, 196
PES (payments for ecosystem services), 205
petroleum-laced materials, 114

PFOS (perfluorooctane sulfonates), 180–81; ban of, 180; and groundwater contamination, 181; at military bases/airports, 180–81
Phoenix, 168–69, 169*fig.*
photographs, 99–100
physical illnesses/injuries, 28; and power lines, 103; and Tunnel Fire (1991), 92, 94, 97–99, 103
Piedmont (Calif.), 193
Pike National Forest (Colo.), 174
Pimlott, Ken, 75, 121
Pincetl, Stephanie, 27
pine forests, 174; and afforestation, 62, 64, 66–67, 69; evapotranspiration rates, 49; and mountain pine beetles, 34, 116–18, 120, 124; pinecones, opening of, 43; and timber extraction, 59; in Tunnel Fire (1991), 3, 135
pittosporum, 5
Pleasanton (Calif.), 80
Plum Creek (Mont.), 203–4
police: loss of life in fire, 154; and Tunnel Fire (1991), 3–5, 154
political capital, 101–2
political discourses: and climate change, 119–120; and eucalyptus, 141; and firestorm as term, 116; and the Incendiary, 110, 116, 119–120, 124–29, 141, 219n1; and wood shingle industry, 124–29
political ecology, 8, 23, 25–28, 34, 88, 92–93
political economies, 29, 37, 40–41, 70; and eucalyptus, 132–33, 139; and the Incendiary, 129, 132–33; and postdisaster reconstruction efforts, 103–5, 166
politics of place, 139–140
ponderosa pine, 117
population growth, 27, 42–43, 45–48, 58, 207; density as predictor of fire activity, 48; and the Incendiary, 114, 119; and reducing demand for development, 205–6; and Yarnell Hill Fire (Arizona), 187
postdisaster reconstruction efforts, 16, 36–37, 45, 80, 149–150, 155–184, 159*maps*, 169*fig.*; and capital improvement recommendations, 36, 149, 155, 167; and Claremont Hotel and Resort, 215n1; and collectivized vulnerability reduction, 36, 91, 101–6, 166; and construction-permitting procedures, 174–75; and estate-based wealth protection, 167, 170; and high-density residential housing, 79*fig.*; and home augmentation, 37, 170–76, 172–73*figs.*;

postdisaster *(continued)*
and individual actions, 105–6; and opportunism, financial, 169–170; and power lines, 103–5, 157–58, 164–67; and private firefighting industry, 179; and roads, 36, 157, 160–64, 162*fig.*, 163*fig.*, 167; and tax revenues, 79*fig.*, 83, 104, 157, 160, 163–64, 166–67, 175–76; and Waldo Canyon Fire (Colo., 2012), 171, 174; and water infrastructure, 80, 103, 155–160, 159*maps*, 165, 167

power lines: aboveground lines, 103–4, 150, 165–66; and arcing phenomenon, 165; downed live power lines, 97, 103, 150, 152, 154, 160, 164–65; and electrically powered water pumps, 158–160, 159*maps*; and exploding transformers, 165; and the Incendiary, 114; and postdisaster reconstruction efforts, 103–5, 157–58, 164–67; and substations, 165; and toxic exposure, 104; and Tunnel Fire (1991), 36, 97, 103–5, 150, 152, 154–55; underground lines, 104–5, 157, 166–67, 194

PRAs (private responsibility areas), 195–98, 195*fig.*, 201

"Prepare, stay and defend, or leave early" (Australian policy), 195, 195*fig.*

preparedness, fire, 23, 132, 143, 160, 190–94, 192–93*table;* and PRAs (private responsibility areas), 195

Prescott (Ariz.), 185, 187

prevention of wildfires, 33, 77, 79–80, 83–85

private property owners/rights, 20, 35–37, 47; and climate change, 121; and do-it-yourself mitigation, 180, 183; and fire preparedness measures, 191; and home augmentation, 37, 170–76, 172–73*figs.;* and the Incendiary, 116, 121, 130; and limiting number of vehicles, 163; and postdisaster reconstruction efforts, 36, 155, 157, 163–64, 166–67, 170; and private firefighting industry, 177–184; and race/class, 82–88, 86*map*, 183; and reducing demand for development, 204–5; and SRA fees, 142–43, 145–46; and tax revenues, 32–33, 71–88, 73*map*, 78*table;* and WPAD fees, 143–46

privatization: of firefighting, 18, 35, 37, 45, 177–184, 194, 202–3; of "war on terror," 170, 177

privilege: and race/class, 81, 83–84; and Tunnel Fire (1991), 34, 90–92, 105–6; and water/power line upgrades, 36, 105, 167

Proceedings of the National Academy of Sciences (PNAS), 118–120

professionals, 97, 99

profits, 18–19, 25, 37, 188–89, 208; and afforestation, 68, 70; and eucalyptus, 133, 139; and the Incendiary, 120, 128, 133, 139; and postdisaster reconstruction efforts, 169–170, 176; and private firefighting industry, 184; and tax revenues, 74, 78, 81; and wood shingle industry, 128

property tax revenue, 32–33, 71–88, 73*map*, 78*table*, 218n18, 218n26; deferrals of, 83; and home augmentation, 175–76; and race/class, 82–88, 86*map;* and reassessments, 72, 84; and Working Forest Project (Mont.), 204

property values, 45; and afforestation, 64, 66, 69, 133; and home augmentation, 37, 170–76, 172–73*figs.;* and the Incendiary, 116, 133; and postdisaster reconstruction efforts, 37, 103–5, 157, 163–64, 166–67, 170–76, 172–73*figs.;* and PRAs (private responsibility areas), 197; and reducing demand for development, 204–6; and tax revenues, 72, 74, 77–78, 85–88, 86*map*. *See also* valued landscapes

Proposition 13, 32–33, 72–75, 73*map*, 82–85, 88, 143, 175, 218n18, 218n26

psychological distress: and the Incendiary, 41; "psychological and emotional maze," 90, 96; and psychological services, 84; and tax revenues, 83; and Tunnel Fire (1991), 33–34, 90–91, 93, 95–97, 101

public lands, 45, 47, 196, 207; and mountain pine beetles, 117–18; national forests/grasslands, 47, 196; public/private land classifications, 23, 27; and reducing demand for development, 203–4; regional parkland, 15

public programs/services, 66, 72, 78, 82, 84

Pyne, Stephen, 46

pyrite, 60

pyrocumulous clouds, 112

race, 81–88, 86*map;* and eucalyptus, 139–42; and PRAs (private responsibility areas), 197; race-based planning decisions, 212n23; race-based violence, 92; and reducing demand for development, 205; and Tunnel Fire (1991), 96, 102; and variabilities of cause, 94

racism, 87–88, 140–41

rail systems, 57; Key Route system, 64; light rail systems, 206; railroad ties, 64; transcontinental railroad, 58
rainfall, 15, 49, 75, 214n28, 220n14; lack of, 111; and mountain pine beetles, 117; reemitting toxins back into air, 97–99
Ray-Bennett, N. S., 28
real estate market, 31–32; and afforestation, 62, 64–69, 65*fig.*, 139; and agent commissions, 176; and Claremont Hotel and Resort, 55, 57, 62; and disclosure requirements, 202; and home augmentation, 176; and the Incendiary, 109, 139, 188; and race/class, 85, 87–88, 140; and reducing demand for development, 202; and road construction, 60, 61*map*, 66; and tax revenues, 74, 85, 87–88. *See also* home construction
Realty Syndicate, 66
red cedar, 125. *See also* cedar shakes/shingles
Red Cedar Shingle and Handsplit Shake Bureau, 126–27
Red Fire (2014, Arcata, Calif.), 49–50
red-flag fire days, 75, 191
redistribution of wealth, 74, 81–82, 88, 218n26; and SRA fees, 143
"Reducing demand for development," 198–207, 198*fig.*
redwood forest, 58–59, 60*fig.*, 66–67; and Tunnel Fire (1991), 2, 5, 215n2; value of, 59, 66
"Refining and expanding community adaptation," 190–94, 192–93*table*, 198
reforestation, 31–32, 68
Reisner, Marc, 27
religious persecutions, 92
relocation after wildfires, 96–97
renters, 82, 84, 88, 97, 204–6; rent-gap value, 204–5
repoliticization, 20, 36, 47–48, 130, 132, 139
restrictive covenants, 190–91
Richmond (Calif.), 193
Riley, James, 154, 164
risk, fire, 8, 18–21, 25, 27–38, 208; and adaptation measures, 191, 194; and afforestation, 35, 67–70; assumed risk, 91–93; and Claremont Hotel and Resort, 57; and climate change, 120–24, 122–23*fig.*; defined, 28; depoliticization of, 36, 58, 110; discouraging development, 198*fig.*; and eucalyptus, 132–34, 139, 141, 221n3; and extraction activities, 58, 60, 68, 215n2; high-risk vs. low-risk areas, 153, 156; imposed risk, 91; and the Incendiary, 31–32, 45, 48, 50–51, 58, 109–10, 112–16, 115*fig.*, 120–21, 130, 132, 188–89, 189*fig.*, 220n9, 221n3; and mobility, 94; and postdisaster reconstruction efforts, 155–56, 161, 176; and PRAs (private responsibility areas), 195–98, 195*fig.*; and private firefighting industry, 178–181, 184; and race/class, 81–85, 88; and red-flag fire days, 75, 191; reduction of, 103–5, 155, 198–203; root causes of, 8, 19–21, 25, 35, 37–38, 58, 81; and SRA fees, 142–43, 145–46; subsidization of, 83; and tax revenues, 75, 77–79, 78*table*, 81–85, 88, 176, 218n26; and Tunnel Fire (1991), 15–16, 33–34, 36, 57, 69–70, 90–96, 101, 103, 105, 153; and variabilities of cause, 94; VHR (very high risk), 176, 191, 195, 198, 198*fig.*, 200–203; and wood shingle industry, 34–35, 69, 124, 128; and WPAD fees, 145; and Yarnell Hill Fire (Arizona), 37–38, 187–88. *See also* vulnerabilities
risk society, 112–13
roads, 32, 71; access roads, 55, 57, 62, 69, 152; arterial roads, 62, 109; blocked roads, 152–55, 161–65, 163*fig.*, 182; and bollards, 163; and bottlenecks, 160–61, 163*fig.*; and cutbacks, 164; and fire preparedness measures, 191; narrow roads, 15, 97, 150, 153–55, 161–64, 162*fig.*, 163*fig.*, 167; and no-parking zones, 162–63; and parking changes, 161–64, 162*fig.*; and postdisaster reconstruction efforts, 36, 157, 160–64, 162*fig.*, 163*fig.*, 167–170, 169*fig.*; and private firefighting industry, 182; and real estate speculation, 65*fig.*, 66–69; spur roads, 66; and tax revenues, 80; and timber extraction, 15, 60–62, 61*map*, 65*fig.*, 216nn7–8; and Tunnel Fire (1991), 36, 97, 150–55; and Yarnell Hill Fire (Arizona), 187
Rockridge district, 87–88, 156–57; Rockridge Apecial Assessment District (RSAD), 157; Rockridge Area Water System Improvements Project, 157
Rocky Mountains, 50, 70, 174; Beetle Incident Management Organization, 118
root causes, 19–21, 25–26, 31, 38, 40–41, 58, 208; and climate change, 120; economic root causes, 20; and Fort McMurray fire (2016), 43; political root causes, 16, 20–21; and race/class, 81; strategies to combat root causes, 189–208

San Antonio (Tex.), 45
San Bernardino County (Calif.), 209n3
San Diego County (Calif.), 124, 209n3
San Francisco (Calif.), 27, 32; and afforestation, 64, 67–68, 70; business class, 64, 67–68; and nozzle hook-up incompatibilities, 158; and population growth, 58; and timber extraction, 58–60. *See also* San Francisco fires
San Francisco Bay Area, 32, 45, 57–59; and afforestation, 67–70; and Diablo winds, 15, 55, 191; and extraction activities, 58–59; and Tunnel Fire (1991), 2, 15. *See also names of cities and regions within the Bay Area*
San Francisco fires, 58–59; Earthquake and Fire (1906), 14, 111, 155, 209n2
San Luis Obispo County, 137
Santa Ana winds, 55
sawdust, 114
sawmills, 59, 60*fig.*
scientific discourses, 16, 20, 29, 36, 208; and climate change, 118–120, 124; and eucalyptus, 141; and firestorm as term, 112–16, 115*fig.*, 129; and the Incendiary, 110, 112–120, 115*fig.*, 129, 141; and mountain pine beetles, 117–18, 124; rapid science vs. slow science, 29–30
seed germination, 43
Self, Robert, 32
senior citizens, 3–4, 94–95, 100, 175
sensitivities, 28, 34, 94–95, 101
"Separation of Sources Act" (Calif.), 74
September 11, 2001, World Trade Center Fire, 170, 177, 209n2
shantytowns, 59
Shock Doctrine (Klein), 170, 176
Sierra Nevada foothills, 59
single women, 96, 102; single women of color, 102
Skagit (Wash.), 204
slums, 207
Smart Growth planning, 46, 198*fig.*, 203
Smith, F. M., 66
smoke, 23, 125; and private firefighting industry, 178; smoke detectors, 145; smoke inhalation, 97; and Tunnel Fire (1991), 57, 149, 151, 153–54, 161, 165
snow, 49, 220n14
social origins of fire risk, 19–21, 26–28; and AVI, 25; social responsibility, 32, 81; and social stratification, 87–88; and Tunnel Fire (1991), 14. *See also* vulnerabilities

socioecological processes, 25, 116, 121–22, 141, 189, 220n6
socioeconomic factors, 19–21; and climate change, 120, 122–23*fig.*; and the Incendiary, 19–20, 120, 122–23*fig.*, 220n6; and tax revenues, 82; and Tunnel Fire (1991), 33; and United Policyholders (UP), 102
Sonoma County (Calif.), 137
Southeast Asia, 170
Southern California, 49, 55, 209nn3–4; and private firefighting industry, 182; and Santa Ana winds, 55
species migration, 136
spot fires, 89–90, 98, 112, 125, 174
SRAs (State Responsibility Areas), 142–43, 145–46
state mitigation funds, 44–45, 83, 121, 175, 188, 190, 193; and reducing demand for development, 200–202
Stengers, Isabelle, 29–30
Stockholm Convention, 180
street lighting systems, 103
stress, emotional/psychological, 28; and Tunnel Fire (1991), 94–97
subalpine forests, 49
suburban landscapes, 15, 18–22, 31–32, 37–38; and afforestation, 67–68; and Claremont Hotel and Resort, 57; and fire preparedness measures, 191; and the Incendiary, 45–49, 48*fig.*, 109–10, 190; and private firefighting industry, 178, 184; and race/class, 82, 85; and reducing demand for development, 198, 203, 205–6; regional suburbanization, 74; and tax revenues, 32–33, 35, 71–74, 73*map*, 76–77, 81–82, 85, 218n26; and Tunnel Fire (1991), 8, 15–16, 57, 90–91; and wood shingle industry, 34–35, 124–25; and Yarnell Hill Fire (Arizona), 187. *See also* development, urban; home construction
succession, ecological, 43, 48
suicides, 95
sulfur, 60
suppression, fire, 23, 43–45, 51, 69, 75, 196; and reducing demand for development, 203–4; and SRA fees, 142, 146; and Tunnel Fire (1991), 151; and Working Forest Project (Mont.), 203–4; and WPAD fees, 145–46
survivor's guilt, 98–99
symptoms, 19, 25, 31, 37–41, 43, 47, 190, 207–8
synoptic analysis, 18, 31, 81, 91–92

INDEX   253

tar sands deposits, 19, 42
Task Force on Emergency Preparedness and Community Restoration, 132
tax revenues, 32–33, 71–88; and home augmentation, 175–76; and the Incendiary, 109; income taxes, 84; and operating budgets, 32, 74–77, 79–85, 157, 160, 163–64, 166–67, 218n18; and postdisaster reconstruction efforts, 79*fig.*, 83, 104, 157, 160, 163–64, 166–67, 175–76; and Proposition 13, 32–33, 72–75, 73*map*, 82–85, 88, 143, 175, 218n18, 218n26; and race/class, 81–88, 86*map;* and reassessed property taxes, 72, 84; sales taxes, 80, 84; and SRA fees, 142–43, 145–46; tax roll analysis (2012), 77, 78*table*, 217n17; and WPAD fees, 143–46. *See also* property tax revenue
telephones, 5
television, 5–6
temperatures, rising, 49–51, 118–19, 122–23*fig.*, 214n28
Texas, 125
3M, 181
thunderstorms, 186
timber extraction, 58–62, 60*fig.*, 68–69, 215n2, 216nn7–8; and clear-cutting, 58, 131
time and space, 21, 25–26, 85, 87–88
tinderbox, 40. *See also* flammability
tourist economies, 170
toxic exposure, 97–99, 104; and PFOS (perfluorooctane sulfonates), 180–81
tragedies: and Tunnel Fire (1991), 92–95, 97, 101, 188; and Yarnell Hill Fire (Arizona), 37, 188
trailer parks, 197
Transforming California: A Political History of Land Use and Development (Pincetl), 27
transportation-oriented developments, 198*fig.*, 206
trauma, 83, 190; and Tunnel Fire (1991), 90, 94–95, 97, 106
tree cover, 15–16, 32, 58, 63*fig.*, 64, 65*fig.*, 67, 69, 71, 87; and eucalyptus, 134; and the Incendiary, 109
tree farms, 62, 65*fig.*
Trust for Public Land, 204
Tunnel Fire (1991), 11–18, 12*map*, 13*map*, 17*map*, 21–22, 28–29, 31–33, 36–37, 196, 208; and afforestation, 66, 68–70; and Claremont Hotel and Resort, 55–58, 56*figs.;* and emergency radio communications, 36, 76–77, 89–90, 149–155, 160, 167; and eucalyptus, 1–3, 132, 134–37, 141; and the Incendiary, 109, 115*fig.*, 188; loss of life in fire, 8, 11, 93–95, 97, 103, 154, 161, 164–65, 196; and mutual aid, 76–77, 80–81, 151, 157–58, 181–82; origin of, 11, 13*map*, 79*fig.*, 137, 159*maps;* personal account of, 1–8, 7*figs.*, 94; and postdisaster reconstruction efforts, 36–37, 45, 80, 91, 101–6, 149–150, 155–167, 159*maps;* and race/class, 83, 85; residents as fire victims of, 1–8, 89–106, 144; and SRA fees, 142; and tax revenues, 75–81, 79*fig.*, 83, 85; and timber extraction, 58, 62, 215n2; and wood shingle industry, 124–26; as "worst-case scenario" wildfire, 14, 155, 187, 209nn2–3. *See also* postdisaster reconstruction efforts

underlying drivers of wildfires, 19–20, 34–35, 38–40, 190, 208; and climate change, 118–124, 122–23*fig.*
understory vegetation, 69; removal of, 131, 135. *See also* litter, ground
undeveloped/unoccupied land, 22–23, 110, 121, 199, 203–7; nondevelopable areas, 204–5. *See also* open spaces
unemployment, 84
unified methodology (FEMA), 131, 137
Union City (Calif.), 80
Union of Concerned Scientists, 50, 122–23*fig.*
Union Pacific Railroad, 64
United Policyholders (UP), 102–3; website of, 102
University of California, Berkeley, 2, 57, 125, 193; and eucalyptus, 131, 134, 221n1; Fire Mitigation Committee, 134
University of Colorado, Boulder, 118–19
urban fires, 14, 85, 155, 209n2
urban peripheries, 15–16, 18–20, 22, 25, 189–190, 207–8; and AVI, 25; and climate change, 121; and the Incendiary, 31, 45, 47, 121; lucrative landscapes at, 18–20, 25; and PRAs (private responsibility areas), 196–97; and private firefighting industry, 184; and reducing demand for development, 198, 205–7; and tax revolt movement, 76–77; and Tunnel Fire (1991), 91. *See also* WUI (wildland-urban interface)
urban/urbanizing environments, 26–29; and AVI, 31; core urban areas, 15, 72, 88, 205–8; in-fill practices, 46, 198, 198*fig.*, 205–7; land conversion, 47, 48*fig.;* and PRAs (private responsibility areas), 196; and reducing demand for development,

urban/urbanizing *(continued)*
198*fig*., 200, 203–5; and tax revenues, 145; urban encroachment, 42; urban growth boundaries, 198*fig*., 203–5; urban renewal, 198*fig*., 205–7; urban sprawl, 35, 45–47, 110–11, 187, 200, 203, 207. *See also* development, urban
urban wildfires, 11, 14, 149, 155. *See also* Tunnel Fire (1991)
Utah, 46, 48*fig*.

valued landscapes: and afforestation, 66–67; and Claremont Hotel and Resort, 57, 62; and tax revenues, 80
variegated vulnerabilities, 21–22, 33–34, 91, 93–101; variabilities of cause, 94; variables of effect, 94
vegetation cover, 68–69, 75; and Claremont Hotel and Resort, 56*figs*., 57; and eucalyptus, 132–36; and reducing demand for development, 200; and Tunnel Fire (1991), 56*figs*., 57, 132–36; understory vegetation, 69; vegetation clearance ordinances, 80, 190–91, 192–93*table;* vegetation clearing, 41, 62, 80, 114, 142–44, 186–87; and WPAD fees, 143–44; and Yarnell Hill Fire (Arizona), 185–87
viewshed preservation, 113, 205
vulnerabilities, 19, 21–22, 25–33, 36, 41–42; and adaptation measures, 194; and afforestation, 35, 68–71; and climate change, 120; collectivized vulnerability reduction, 36, 91, 101–6, 167; defined, 28; embodied, 90; and eucalyptus, 132; and extraction activities, 31–32, 58; and firefighters, 90; and the Incendiary, 47, 69–70, 112, 120, 132, 188, 189*fig*.; net vulnerability, 83; and PRAs (private responsibility areas), 195*fig*., 197; and race/class, 81–83, 85, 87–88; and risk society, 112; root causes of, 19, 25–26, 120; and SRA fees, 145; and tax revenues, 32–33, 75, 81–83, 85, 87–88; and Tunnel Fire (1991), 2, 16, 21–22, 33–34, 36, 83, 89–90, 106; variegated vulnerabilities, 21–22, 33–34, 91, 93–101, 197; "vulnerability-in-production," 128–29; and wood shingle industry, 128–29; and WPAD fees, 145; and Yarnell Hill Fire (Arizona), 188. *See also* AVI (affluence-vulnerability interface); risk; fire

Waldo Canyon Fire (Colo., 2012), 171, 174–76
Walker, Dick, 87

"war on terror," 170, 177
Washington (State), 46–47, 48*fig*.
water infrastructure: emergency backup generators, 80, 160; and gravity flow, 158; historical pipe installations, 156; hydrants, 76, 89, 152–53, 156–58; and loss of electricity to water pumps, 158–160, 159*maps*, 165; and nozzle hook-up incompatibilities, 76, 157–58; and population growth, 58; and portable pumping units, 160; and postdisaster reconstruction efforts, 80, 103, 155–160, 159*maps*, 165, 167; reservoirs, 80, 158–160, 159*maps;* and Tunnel Fire (1991), 4, 11, 36, 76, 89–90, 97, 150, 152–53; water pipe upgrades, 156–57, 167, 194; water pressure, 4, 150, 152–53, 156–58; water pumps, 80, 103, 152, 158–160, 159*maps*, 165; water shortages, 196; water supply, 97, 155–56, 160; water tanks, 89–90, 103, 152–53, 165
water purification, 205
Watson, Diane, 124, 127
Watts, Michael, 28
wealth. *See* affluence
welfare, 84
wheat, 59
whitebark pine, 117
whites, 32, 46, 72, 87–88, 134, 140–41
Wildfire Defense Systems, 178
wildlands, 14–15, 22–23, 47; wildland-urban intermix, 22–23, 24*fig*.. *See also* WUI (wildland-urban interface)
willows, 137
winds: and afforestation, 64, 69; Diablo winds, 15, 55, 191; and eucalyptus, 134–35; and firestorm as term, 112–13; Santa Ana winds, 55; and Tunnel Fire (1991), 14–15, 55, 57, 89, 110, 135, 165; and Waldo Canyon Fire (Colo., 2012), 174; and Yarnell Hill Fire (Arizona), 185–86
winter die-off of insects, 117
winter wildfires, 49–50, 75, 220n14
win-win outcomes, 36, 150, 155–56, 167
Wisner, Ben, 26, 28
Witch Fire (San Diego County, Calif., 2007), 209n3
Wood, Samuel, 27
wood fuel, 64
wood shingle industry, 34–35, 69, 124–29, 134; and cedar shakes/shingles, 34–35, 124–29; and fire-retardant/fire-resistant shingles, 124–28, 221n24; and the

Incendiary, 34–35, 114, 124–29, 134; lobbying by, 124, 126–28; and Tunnel Fire (1991), 1, 4–6
Working Forest Project (Mont.), 203–4
World War II, 68, 70–71; and firestorm as term, 111; and Hiroshima, 93, 111
World War II, post, 74; and tax revenues, 81, 84, 88
WPAD (Wildfire Prevention Assessment District) fee, 143–46; rejection of, 144–45
WUI (wildland-urban interface), 14–15, 18–20, 22–26, 24*figs.*, 30–31, 37–38, 40, 207–8; critics of, 46; and the Incendiary, 34–35, 46–47, 48*fig.*, 111, 140*fig.*, 188, 190; and land conversion, 47, 48*fig.*; and postdisaster reconstruction efforts, 171, 174; and PRAs (private responsibility areas), 197; and private firefighting industry, 177; and reducing demand for development, 198*fig.*, 200–202; shift away from, 23, 25; and Waldo Canyon Fire (Colo., 2012), 171, 174; and wood shingle industry, 128; and Yarnell Hill Fire (Arizona), 186
Wyoming, 48*fig.*

xenophobia, 140

Yarnell (Ariz.), 185, 187
Yarnell Hill Fire (Arizona), 37, 185–88, 196; and active defense of communities, 187–88; loss of life in fire, 187–88; origin of, 185
Yavapai County (Ariz.), 187; Community Wildfire Protection Plan, 187

zoning ordinances, 190–91, 198, 198*fig.*, 200, 203–5, 207